Economic Inequality—Trends, Traps and Trade-offs

Published 2021 by River Publishers
River Publishers
Alsbjergvej 10, 9260 Gistrup, Denmark
www.riverpublishers.com

Distributed exclusively by Routledge
4 Park Square, Milton Park, Abingdon, Oxon OX14 4RN
605 Third Avenue, New York, NY 10158

First published in paperback 2024

Economic Inequality—Trends, Traps and Trade-offs / by Medani P. Bhandari, Shvindina Hanna.

Routledge is an imprint of the Taylor & Francis Group, an informa business

ISBN: 978-87-7022-391-1 (hbk)
ISBN: 978-87-7004-305-2 (pbk)
ISBN: 978-1-003-33800-0 (ebk)

DOI: 10.1201/9781003338000

Contents

Dedication

Firstly, we dedicate this book to all loved ones Worldwide who lost their lives due to Corona pandemic from the beginning to date; we acknowledge that some of you were victimized due to inequality in healthcare systems. In your memory—in this book, we have raised the questions of why the world was not able to save you. The world will remember and salute you all, and we hope that future science will be able to control such pandemic.

And to

Prof. Oleg Fedorovich Balatskyi (May 11, 1937, Ulan-Ude, now Buryatia, Russia - June 29, 2012, Sumy, Ukraine)

You were, are and will be remembered for your contribution for societal change through Knowledge, Wisdom, and Morality.

Oleg Fedorovich Balatskyi (May 11, 1937, Ulan-Ude, now Buryatia, Russia—June 29, 2012, Sumy, Ukraine) is a well-known Ukrainian scientist in the sphere of Environmental Economics. He gained academic degrees of Doctor of Economics (1980), Professor (1981), Honored Member of Science and Technology Association (1992), Academician of the Engineering Academy of Ukraine (1992); Academician of the Ukrainian Ecological Academy of Sciences (1992), member of the Society for Nature Protection (1975–1990), Professor of management at Sumy State University, Honored Professor of Sumy State University (1992), founder of the Scientific School of Environmental Economics in the region.

Oleg Balatskyi was an innovator and founder of the national school of Environmental Economics that has titled "Economics of Natural Resources and Environmental Protection" at the beginning.

He was an idea bringer, and he was the first scholar who implemented a theoretical and practical basis for the compensation payments for the damage to the environment. The economic losses caused by air pollution were estimated for the first time in the Soviet Union in 1989 as an experiment, and then the approach was approved at the legislation level in 1992 in Ukraine.

The main results of Balatskyi's research were published in more than 350 scientific articles, book chapters, books, textbooks, guidelines, etc. His works were seminal for the development of the academic stream and are exemplary in terms of academic writing style, argument construction, and empirical data processing. His fundamental monograph "Clean Air Economics" (1979) became the methodological basis for the development of environmental economics in Ukraine and post-soviet countries. As a result, the international collaboration joint Ukrainian–Belgian textbook for higher education "Economics of Nature" was published under Professor Balatskyi's edition. The textbook has been published in Russian and English languages.

Professor Balatskyi was famous for his creativity in managing complex systems, punctuality in holding and moderating meetings. Working within the deadlines was another strength; he was one of the best at managing large-scale projects and team building. As a Head of the department, he became a mentor for many scholars in the field who are now famous and found their own scientific schools. Professor Balatskyi was known by his exceptional leadership style; his statements became famous and are now used as quotes in management.

Words from his students

Professor Balatskyi passed away in 2012, and he was a man of excellent service, a distinguished person in science and techniques, the Academician of Engineering Academy of Ukraine, Doctor of Science, Academician of

International Academy of Safety. Professor Balatskyi was one of the founders of the Ukrainian school of Environmental Economics and a founder of the School of Economics in the Sumy region. This school became an alma mater for tens of Doctor of Sciences and hundreds of Ph.D.s. More than 300 scientific works were published, and among them, more than 20 monographs and books became a basis for several generations of scholars in the field. He launched a new academic direction of studies, and today the theory of evaluation of economic losses due to environmental damage is still relevant.

Oleg Fedorovich was an outstanding manager of scientific research projects. To have more than ten R&D projects in a process simultaneously was typical for his team. The deadlines and performance system were unique and effective; all the projects were accomplished on time.

Oleg Fedorovich always was full of new ideas and focused on a search for new directions, such as economic tools for environmental management, friendly-environment economic policy for the regions, theory, and implementation of goal-oriented complex programs in environment protection. Other streams of his research were the economic potential of administrative and industrial systems; and charging system for using natural resources that was transformed from an experiment in Sumy to the implementation at the government level. It is also worth giving special recognition to the series of work in resources recycling, water in particular. This project was partly implemented at Sumy Production Association (PA) "Khimprom."

As a talented scholar and effective manager, Professor Balatskyi put attention to the staffing of research and teaching, and first, to the recruitment of young scientists. He personally was a supervisor for 15 Doctor of Sciences and more than 50 Ph.D.s. Under his supervision, his three apprentices (Melnyk L., Kozin Y., Boronos V.) performed series of works and won the Academician Proskura Award and then Leninskyi Komsomol Award (one of the most distinguished awards at that time).

Professor Balatskyi initiated the Special Academic Council in "Economics of Nature Resources and Protection of Environment," and that was the beginning of the training of Ph.D.s in the field. He was a member of many scientific journals, and editor-in-chief of Visnyk of Sumy State University, Economy series. Under his influence, section of economics, management and organization became one of the most powerful institutes of Sumy State University. A lot of awards, medals, and victories distinguished the path of Professor Balatskyi.

Oleg Fedorovitch crafted many aphorisms about life, managing people, and leadership. One of them is quoted a lot by the followers: "The past is not under our control, so we should change the future".

Arkadii Zhulavskiy, Professor, Ph.D., Professor at Oleg Balatskyi Department of Management

I've met many good teachers in my life, and Oleg Fedorovich played a unique role. Everything had started from my speech at the conference when I was a student. At that time, this conference was held by the Head of the Department of Management, Professor Balatskyi. For me, being a student, that was the first experience of public speaking and scientific presentation in my research outcomes, first participation in R&D, first publication… From the first meeting, Oleg Fedorovich caught us by his knowledge and expertise, wisdom, and charm.

And then fateful stages happened. When he was hiring me, he asked numerous questions, like "What is your sports category?" or "Do you have siblings?" and then he made a job offer. And of course, I had many offers due to his recommendations, such as scholarship for the PhD training, cooperation with Sumy Oblast Department of Ukrainian Ecological Academy of Science, senior positions at Sumy State University, many international conferences and symposiums, business-trips, teaching practice and even acquaintance with my future wife … All these events were linked with his kind support. The first impression about him became the right choice of my life path—this is a substantial person, Leader, and Wise Mentor, and Coach who you should follow and with whom you should go through life with.

It turns out that my wife and I walk every day on the street named after Oleg Balatskyi, and it has a meaning for both of us as he showed us the way in life, as a professional, and personal as well … My memories about Oleg Fedorovich Balatskyi may be very simple. This is gratefulness and appreciation for the knowledge, the science of life, his goodwill, and kind advice. I am still impressed by the teacher's wisdom, his tolerant strictness, the abilities to motivate people for personal development and encourage them to grow professionally.

Vadym Lukianykhin, Associate Professor of Oleg Balatskyi Department of Management

Professor Balatkyi was an outstanding person who changed the life of his co-workers and team members softly with tolerance and wisdom he gained through many years of managing experience. As one of his Ph.D. students, I was fortunate to know him as a teacher and the head of my department. Professor Balatskyi brought the vivid combination of vividness, sincerity, and perspicacity in the communications with people around him. I got to know him at the end of his career, when he didn't waste much energy on operational challenges, but he was always full of insights for the strategizing and ready to share them. Always generous in terms of time spent on reading your publication and making a kind comment on it. He was a true leader, mentor, and a man of prodigious management talent.

Many famous quotes of him went viral. One of my favorites is "all bad comes naturally, and good has to be organized." Other ones are "those who are not able to obey, are not able to manage," "past is not under our control, and we must strive to change the future." I believe that I am a change maker, and we train new generations of change-makers in Management Studies, derive from Professor Balatskyi's principles and wisdom.

Hanna Shvindina, Dr. Sci., Ph.D., Head of Oleg Balatskyi Department of Management

Preface

"Economic Inequality—Trends, Traps and Trade-offs" presents the unexplored issues of economic inequality, with the exemplary case studies of various countries. Inequality is a chronic divisive factor of society. It is a known truth/fact that several inequalities (economic, social, cultural, religious, geographical, etc.) have been omnipresent in human society. Inequalities can be found within each family, from family to community, and from communities to the nation, and to other nations globally. Inequality is the major cause of political, economic, and social instability that creates crisis and conflict within the society, from societies to nation, and to other nations. The major causes of inequality are unequal, uneven, biased, power-centric distributions of human, economic, social, political, cultural, and spiritual necessities. The human civilizations are grounded in a way that, small elite groups create webs and networks within which they lead the society, rule the society in a way that makes public rarely capable enough to get opportunities even to participate in fair competitions. Because in marginalized societies, individuals would not have similar access to wealth, education, health, and other benefits as the ruler elites have. Therefore, the elite groups remain elite for generations, and the dominated groups remain as victims. This scenario is political inequality, which is the major cause of inequality in the society. The dominant group capture the political opportunities and power of governance so that marginalized groups always remain at the lower strata of the political, social, and economic system. To minimize economic inequality; it is necessary to break the webs created by the traditionally inherited elites as well as by the newly emerged haves group; and empower, encourage, insist, and uplift the have-nots group, in a way that the dominated, marginalized group will also get equal opportunities to survive, grow, and enjoy their lives.

Book begins with the theories behind the inequality and unveils the problems and consequences including the efforts of the United Nations' proposed and implemented Sustainable Development and Sustainable development Goals, which addresses the inequality and examines the Threat to The Sustainable Development and Toolbox of Sustainable Territorial Development. Broadly, the book elaborates Economic Inequality with the theoretical lenses, compares Economic and Social Inequality, unveils the Problems and Consequences of Inequality, shows how Unified Index for Balanced Development relates with inequality, and provides the Inequalities Level Estimation with the country-specific case. Further, book chapters examine inequality with the cases of Territorial Development, provide distinction of Poverty and Inequality, by analyzing Income/financial Inequality—with Poverty Status, Distribution of Income, Relationship Between Inequality and Socio-Economic Drivers, and Green Growth and Entrepreneurship. Most importantly, book chapters questions how inequality is grounded in the society throughout the history, presents current situation, and forecasts the future scenario of Inequality. Book chapters also show the impact of economic inequality, which divides the society in a way that rich continue to be richer and poor continues to be poorer. Book also presents the positive and negative aspects of Liberal Socialism. The year 2020 was under the COVID-19 pandemic regime. The chapter of the book also addresses how COVID-19 pandemic regime has been widening the inequality globally and specifically the developing world, with the proposal of new theory based on the situational analysis of COVID-19 pandemic regime "CORONOMICS."

Further, book examines the major parameters of intertwined socio-economic issues of inequality and focuses on the key economic issues of inequality, namely, income and wealth distribution, equity & equality of outcome, and equality of opportunities. Basically, economic inequality is measured by wealth, income disproportions in distribution and consumption patterns in the specific area. Mostly, for the statistical presentation inequality is measured using Gini Coefficient, inequality adjusted human development index, 20:20 ratio, Palma ratio, Hoover index, Galt score, Coefficient of variation, Theil index, wage share, etc. However, it is not necessary that all income can be measured in the monetary unit and can be presented and proved through statistical tools. In this book, authors tried to unveil the problems and consequences of socio-economic inequality through case studies of different countries (Nepal, USA, India, Mexico, Ukraine, European Union, Russia, Pakistan, Serbia, Iran, Nigeria, Bulgaria, Finland, Middle East, Latin America, North America, Asia, Pacific, etc.). Based on various case studies, this book encourages to rethink economic development through the lens of growing inequalities and disparities. These disparities became obvious with sharing information about global changes and distribution of environmental, economic, and social risks among the recipients. The economic growth per se is disproportional, and the efforts of scholars, practitioners, and policymakers should be directed to empower the marginalized society in a way that no one should get left behind (UN Slogan). The main challenges and focus of every chapter are different, and yet in combination, they give an integrated understanding of the phenomenon of economic aspects of inequality. Primarily, this book can help to understand the problems and consequences of inequality. Therefore, the book can be useful to all concerned stakeholders of society, who are willing to see the undivided, inclusive, and social justice in the society. This book will also be useful to policymakers, social and environmental activists, agencies, educators, and practitioners in the sphere of economics of inequality. The methodology of the research can be replicated and taken forward by future researchers in the field.

Prof. Medani P. Bhandari, PhD,
Akamai University, Hawaii, USA;
Sumy State University, Ukraine

Forewords

Inequality is a popular subject in the contemporary world, and it is **the significant problem of humanity**. Much of the policies and papers have been generated for a long time but the problem of global inequality is increasing in larger areas of the globe fueled by job insecurity, political disruption, large-scale migration, and frustration of the people. Inequality is also connected with wider global issues such as climate change and its impacts, technological disruption, access to technology and knowledge, large-scale urbanization in every parts of the world and yet humanity has not been able to address such mega problems in a judicious way. Governments are the key to make the decision of the countries for caring and sharing the benefits to poor strata of the people. However, in this globalized world their policies and decisions are influenced by mega corporate agencies and other countries' policies and strategies which makes them very difficult to address the agenda of equity. Thus, disparities of various kinds and lack of opportunities to the people are at a rise.

This book **"Economic inequality—Trends, Traps and Tradeoffs"** brings case studies on the subject from different countries. Societies are made up of different groups where the dominating group usually captures and controls the power and political opportunities pushing the marginal group further down. To correct this, there is a need to shift in power structure and upliftment of the poor people. The fundamental for this is the feeling of moral value and willingness to improve the Gini coefficient of each country. National as well as global leaders of different spheres of societies including political leadership must realize and enlighten themselves to circumvent any form of cartelling and matchmaking in running the country or business for sharing and caring of the people in an equitable way. This edited version of the book has examined the major parameters of **intertwined socioeconomic issues** using different inequality parameters including Gini coefficient. This book, I hope will encourage the planners and policy makers and decision makers to think carefully on the subject and suggests on how to narrow down the inequality.

I congratulate to all the contributing authors of the book and to the editors Profs. Medani P. Bhandari and Shvindina Hanna for their arduous work to bring the knowledge to the public.

Dr. Krishna Prasad Oli, Ph.D
Member
National Planning commission
Government of Nepal.

Economic Inequality—Trends, Traps and Trade-offs

The issues related to economic inequality and related matters of discrimination are at the forefront of economic crises worldwide. This book, Economic Inequality—Trends, Traps and Trade-offs, edited by Dr Medani Bhandari and Hanna Shvindina, addresses the importance of universal equality and human rights worldwide.

Important to state here is the reality that the United States of America was founded upon the premise that "*all men are created equal, that they are endowed by their Creator with certain unalienable rights, that among these are life, liberty and the pursuit of happiness.*" (Thomas Jefferson, 1776)

The present-day deterioration of this foundation worldwide discloses the importance of this book in support of our understanding of the path forward away from economic inequality due to the discrimination based on race, religion, ethnicity, gender, LGBTQ+ and traditional male-centric culture, and social, political, economic, and religious systems. It is known that the poverty and deprivation are closely related to deprivation and lack of opportunity that emerge from broad-based inequalities and severe lack of sustainability of cultural advances across the globe. Globally, the richer people are getting richer, and poor are getting poorer. There are no justifiable arguments, why the marginalized people are not getting the opportunities to sustain themselves and be able to compete with the established economic system. This book provides the picture of how society is divided between haves and have-nots, what are the hurdles to compete with the rich who directly or indirectly help to widen the inequality.

Consequently, Bhandari and Shvindina have successfully endeavored to clarify the importance and the path toward reducing pervasive inequalities with the purpose of moving toward minimizing the economic inequality across the global community, by providing the global, regional, and country-specific cases. Our cultures are judged by how we treat the least among us, how we support those who are most abused and what help we provide to those who face hunger and sufferings. I am encouraged by the depth and courage of this book. This book is a depth study of global economic inequality supported by the various specific case studies around the world. This book "Economic Inequality—Trends, Traps and Trade-offs" will be very useful to all stakeholders who care the humanity and care the world.

Prof. Douglass Lee Capogrossi, PhD
President Emeritus, Akamai University, USA

Economic Inequality—Trends, Traps and Trade-offs"

The inequalities illustrated in this book captures the path I have walked in Peru, South Africa, Australia, New Zealand, Canada and yes, even the USA. I have walked into huts, sat among the native people in these countries and witnessed the disparity. I struggled how to communicate, to be accepted, to be accepted and to listen with my whole being. I did not learn these strategies in my bachelor, master, or doctoral education. I learned these skills by walking among and with the people in their native communities firsthand by experiencing their life. From this base, I was changed. Changed to spread what I learned and experienced to others, capturing the knowledge and skill from those who had different experiences from what I had. How can I honor their traditions and mine to come up with a "we and us" approach to life to laugh, have fun, create, explore as equals rather than "I am different," "I am better," "My way is right and yours wrong."

When I started as a psychotherapist, I told my friends the stories of the "good guys" and the "bad guys." My religion and beliefs were the "good guys" and those do not like me were the "bad guys." As I moved among states in the USA, alone I went from a good guy to bad guy, to even more bad guy, and to a good guy when I moved to St. John's Newfoundland Canada and then to North Carolina, once again, not one of the good guys. It doesn't matter if I share my specifics of what the good or bad guy definitions were / are as they change from person to person, from culture to culture. What matters is we separate from each other and devalue each other rather than learning from each other and finding win–win solutions for all.

Learn to experience new food, new ways to build and live in houses. Ways to groom, to heal, and have fun.

Experiment with new learnings and let go of what does not fit for you while taking on what you like without devaluing another. That is the key. One is not better than the other. One is merely different. Nothing more, nothing less.

Simplistic living is made up of basic values of sharing vs. comparing, of uniting vs. dividing, of bending with the breeze rather than standing firm and tall that perhaps can lead to breaks when the wind blows hard and out of control.

Am I speaking too simply with the complex issues of economic inequality? I suspect now as I am now able to flow with the breeze, taste new and come together to solve complex problems, listening, hearing, and speaking my truth and allowing the same from those I am in contact with.

I may be the president of Akamai University now, but as a 76-year-old, I am a mother, friend, sister, grandma, gardener, animal companion, and gate keeper of ancient secrets revealed by being an open channel to Divine energies when inspiration from above "pops" into my head rather than being ego driven. Yes, I accept the fact that I hear voices and have been since a teenager. I follow Divine inspiration and teach and help others to open to their Divine inspiration according to their terms, not ones imposed by me or my guidance. So, as you can tell, I am eager to embark on investigating the knowledge put forth in Medani and Shvindina's book—not separating them in different ranks and genders that should be more acknowledged one from the other, but as equals each on their own mountain top with a different view seen by their eyes with great insight to be pondered, tasted, and ingested as one is ready. And yes, the term is, "is ready." Some of the "truths" I was taught as a young nurse became absolute as I grew in age and experience. Oh my, what was the truth of one day is not now "the way to do things!"

My point has been made I believe, exploring the three T's—trends, traps and trade-offs. My point actually doesn't matter really very much. Only things that matters is did I "tweak" your interest to read on—to find your truth, to counterbalance inequalities that are part of your life's journey. Start at home, in your house, in your community, in your country ... walk the path that unfolds for you and do your best not looking left or right, up or down, forward or backward. No judging or accepting of being judged and not measuring up in some way by somebody. Simply be you. That is good enough. And you are perfect, exactly the way you are.

Thanks, Medani and Shvindina for being you and once again ... polishing the tarnished silver in our ideas to bring out the beauty that sometimes can get hidden from the wear and tear of life.

Dr. Mary Jo Bulbrook,
President,
Akamai University, USA

Economic Inequality— Trends, Traps and Trade-offs

Economic inequality is a complex issue relating to unequal distribution of income, wealth, and opportunity between different groups in a society. This phenomenon is deeply rooted in historical, cultural, political, and economic processes. Living conditions in different places or different countries vary, and the differences are often staggering. Such discrepancies are believed to be the consequence of two centuries of unequal progress. According to Forbes, on 18 March 2020 there were 2.095 billionaires around the globe, while the data from a report published by the World Bank in 2015 quote that in that year about 736 million people were living on less than $1.90 per day. However, it should be emphasized that when compared with 1990, when 1.9 billion people lived in extreme poverty and the world had only 269 billionaires, the situation has improved considerably (Forbes, *The World's Billionaires: 25th Anniversary Timeline,* 2012).

Inequality is a complex issue felt by every modern society. Inequalities are found in the levels of income and consumption, access to information, employment, culture and the degree of participation in public life. As a consequence, large groups of people suffer economic and social exclusion which, in turn, triggers changes in values of a given society. There is a view in business theories that economic and social inequalities are an integral part of the functioning of the economy and the society. The literature on the subject indicates that the basis of economic inequalities that may turn into social inequalities lie in the personal characteristics of individuals and their tendency to compete for their property status.

The issue of inequality is often narrowed down to the economic sphere, in particular to the satisfaction of needs, which is significantly influenced by the rate of economic growth. Although economic growth is necessary to satisfy the needs to a greater extent, it does not translate directly into the equalization of social inequalities. There are various methods of measuring economic inequality, however the reason for conducting the measurements is always the same—the existence of diversified access to material goods leading to the economic inequalities within a given socio-political system, which to a large extent impacts the functioning of individuals and influences the quality of their lives. Economic inequalities and social inequalities are closely connected and interact with each other.

Until recently, in political discourses of countries, social inequalities and economic inequalities were perceived as not very important issues. One of the reasons for this perception of the problem was the belief that social development and the dynamically following civilization progress could lead to a spontaneous limitation of this problem in the near future. In the 21st century, social inequalities are being addressed in the form of policies of public institutions, efforts are concentrated on their elimination and the topic itself frequently occurs in public debate in various countries of the world.

The book by Professor Medani P. Bhandari and Associate Prof. Shvindina Hanna refers to the timeless issues related to economic inequality, sustainable development goals, financialization as a determinant of income inequality, poverty, and discrepancies in access to housing and poverty status of households. The authors of the book attempt to answer the question of what needs to be done and what tools should be used to reduce the scale of economic inequality in the world. The book illustrates in a comprehensive way how the world faces economic, social, and environmental challenges. This book can be recommended for those readers who are interested in the subject of economic inequality and/or for those involved in the elimination of strong economic and social imbalances.

One could hardly disagree with the assertion of the authors that it is largely our responsibility to compensate for the economic and social inequalities between people no matter where they were born, in order to provide them with a decent standard of living.

Dr. Jacek P. Binda, Rector and a researcher at Bielsko-Biała School of Finance and Law, Poland (pol. Wyższa Szkoła Finansów i Prawa w Bielsku-Białej, Polska).

The book "**Economic Inequality—Trends, Traps and Trade-offs**" edited by Prof. Medani Bhandari and Prof. Shvindina Hanna is an important contribution in the study of comparative economic inequality in the world. I am pleased to write a few words about this timely book adding my personal thoughts.

In spite of an historically impressive level of economic progress in the world, the growing economic inequality has remained a stubborn problem for the humanity. According to data from the World Bank, the average global per capita income in 2020 was around $12,500 in current US dollars. While this is an excellent level of income, many poor countries still have very low per capita incomes. For example, the World Bank data for 2019, Somalia had a per capita income of $126 and Burundi $261. Dozens of other countries have annual per capita income of less than $1,000.

While the world is witnessing an unprecedented level of prosperity where billions of people are enjoying a high standard of living, poverty is also equally pervasive. The dramatic levels of wealth equalities can be seen in Oxfam International's report of 2020. The report notes: "There were 2,153 billionaires in 2019, and together they have the same amount of wealth as of the poorest 4.6 billion people in the world." The World Bank estimates that in 2015, almost 690 million people globally lived on less than $1.90/day, the updated measure of extreme poverty.

The economic inequalities often go hand in hand with racial, and ethnic class, and sexual orientation or gender differences. Women generally make less than men for the same work. A UN data for 2020 shows that women still make only 80 cents for every dollar that men make for the same job. The data further indicates that women with children, and minority and immigrant women make even less.

Growing up in the remote hills of Nepal, I have experienced first-hand the impact of poverty on human well-being. In my experience, the hand-to-mouth existence and lack of any opportunities for upward mobility left people in a state of perpetual poverty and despair. The situation impacts human dignity, health and leaves an individual in a state of hopelessness. Today, the economic situation in Nepal has significantly improved, but inequality of income and wealth remain persistent making it difficult for millions of people to live a reasonable and decent life.

Even in wealthy countries like the United States, there are millions of poor people struggling to make ends meet while some 6 percent of the population are millionaires. Similar stories of inequality can be seen in OECD countries. In many developing countries, the inequality situation is even worse.

Lifting people out of poverty and reducing extreme income and wealth inequalities is important not only to provide for the dignity, well-being, and welfare of everyone in the society, but also to ensure that the development remains sustainable. Inequality breeds conflicts, inefficiencies, and lack of security for all. It also is a moral hazard for the entire society.

The book editors Prof. Medani Bhandari and Prof. Shvindina Hanna have extensive background in this field as educators and practitioners in social, environmental, and economic development. Bhandari started his education and a career in the environmental and development field in a rural area in Nepal. He later moved to the USA to obtain his doctorate degree and now teaches in several countries. His experience in different countries have provided him with a comparative perspective on how societies function and how social inequalities can impact people in their abilities to realize the full potential as members of the society. Hanna grew up and educated in Ukraine. She later spent time in France and USA for higher education. Her education and training in economics, management, and development has helped her to teach and write extensively in support of social and economic development internationally. Her vast education and experience have provided her with important perspectives to co-edit this book.

The editors have presented an excellent collection of chapters by prominent authors from many countries who provide a rich variety of stories from their experience, study, and research. As a result, the book is rich in comparative perspectives and analyses providing the readers with a wide view on the topic of economic inequality.

This book will be an important reading for students, academics, researchers, policymakers, and laypeople who are interested to understand economic inequalities and learn about possible policies and programs to reduce inequality.

Dr. Ambika P. Adhikari, PhD.
Phoenix, Arizona, USA
Sr. Sustainability Scientist
Arizona State University, Arizona, USA
February 2021

Economic Inequality—Trends, Traps and Trade-offs

A simple but very powerful saying "money attracts money" works at all levels whether it is at individual, household, community or at the national level when it comes to wealth generation and distribution. As wealth gets generated to those who already have wealth in hand and does not get generated to those who lack wealth in their hands, this ever-increasing divide between haves and have-nots in every societies globally is inevitable unless appropriate policy measures for wealth generation and distribution in societies are implemented and guaranteed. Economic inequality is a widespread global problem. It is associated with so many social inequalities including access to good education, healthcare facilities, food and nutrition, decent housing, leisure and vacations, and old-age security. As large amount of wealth is being accumulated to a relatively small proportion of individuals in a society leaving majority of the people struggling in meeting their basic needs of food supply, housing, health, and education, many societies across the globe are experiencing political unrest, social disorders, riots, and environmental degradation. Many activities related to environmental degradation such as deforestation and land degradation especially in developing countries are associated with increased population and poverty. To bring societies out from the vicious cycle of poverty and economic inequality, countries across the world need to design and implement sustainable economic development measures that address both sustainable income generation and the distribution of wealth. Economic equality leads to the creation of socially just and sustainable global societies.

This book "Economic Inequality—Trends, Traps and Trade-offs" is a treatise of facts and knowledge on economic inequality. This book includes chapters written by well-established and renowned scholars in the field from across the globe. Book chapters give full insights into the theoretical and practical aspects of economic inequality including a balanced and sustainable development and green economy. The causes and consequences of economic inequality in the societies across the globe are well analyzed and presented. The authors have done an excellent job in filling the knowledge gap on economic inequality. This book will be a very valuable resource to those who like to understand and learn about economic inequality and want to be an agent of change for creating an economically and socially just global society.

Prof. Dr. Durga D. Poudel,
University of Louisiana at Lafayette, Lafayette, Louisiana, USA

Economic Inequality—Trends, Traps and Trade-offs

Profs. Medani P. Bhandari and Shvindina Hanna which contents several chapters of the book entitled ***Economic Inequality—Trends, Traps and Trade-offs*** explore various causes and consequences of income inequality. Their analyses start from liberal socialism, balanced sustainable development, and green investment. They dissect the roles of various determinants of income inequality, household level poverty status using examples from urban and rural areas and cross countries' information, and debt-ridden inequality in the north and global south. They also present some means of forecasting income inequality using tools like Gini coefficient, inequality adjusted human development index, 20:20 ratio, Palma ratio, Hoover index, Galt score, Coefficient of variation, Theil index, and wage share. With careful analyses, the authors have challenged the planners, policy makers, economists, social scientists, and politicians (commonly referred as "development actors") to link the current income inequalities with the location-specific facts and analyze such location-specific geographic causes leading to income inequality to the global level if at all the "development actors" are seriously thinking of alleviating multidimensional inequality on a sustainable basis.

Inequality exists everywhere and at every level, and becomes the major cause of political, economic, social, cultural, and religious rifts, and instabilities and creates crisis and conflict within and between society and nations. Inequality started due to skewed distributions of various resources from traditional society, and it has been widened in various ways in the modern society. Economic inequality has brought stratifications in all aspects of life from living conditions to educational levels. The authors have analyzed how social, cultural, ethnic, gender, and humans' levels of thinking have created inequality in a society. The authors have critically reviewed many literatures and have concluded that how despite of the many efforts made by the United Nations (2015) including its Goal 10 to alleviate income inequality, economic inequality is increasing from a societal level to global level. Many international organizations, governments, non-governmental organizations, and other agencies have played significant roles in reducing inequality, yet inequality prevails in all aspects of local and global society.

The authors suggest that income inequality can be minimized through the empowerment of low-income and "have nots" groups from the local to national and international levels and through the processes of economic convergence. The authors warn that if present trends of income inequality continue, at the global level, the share of income would be concentrated in the hands of a wealthy few, while the share of income at the bottom will be decreased further and the pyramid of income inequality will sharply taper instead of tending toward a plateau. From their review of literature, the authors seem to have favored the European model to narrow down the gap between the rich and poor through progressive taxing methods. The authors warn that the income inequality reduction is not possible merely by decree and economic growth, but it might be possible through sincere collaborative efforts and gender parity and scientific taxing methods in both developing and developed countries. To justify their arguments, the authors present several location-specific facts from rural and urban aspects from different countries and deeply analyzed the demographic, ethnic, religious, and cultural perspectives and suggest measures that can be taken to minimize income inequality.

This is a very useful book for "development actors" and educators. The authors deserve congratulations for this monumental work.

Prof. Keshav Bhattarai, Ph.D.
Professor of Geography
School of Environmental, Physical & Applied Sciences
University of Central Missouri,
Humphreys 223C, Warrensburg, MO 64093
bhattarai@ucmo.edu

Economic inequality is a chronic economic inter-generational disease of society and mother of all inequalities. It must, therefore, be treated as a global pandemic and declare war against this universal economic scourge. This suffering has exposed deep socio-economic divide at the forefront during the pandemic and revealed serious weaknesses in modern economic strategies and, of systems. Exacerbating the situation worldwide by spread of contagious Coronavirus, societal vulnerabilities have multiplied fragility of humanity as it touched every family, individual, society, and nation, in one way or other, and destroyed past mitigating efforts. Exposed fault lines were largely shrouded within the sphere of societal level, race, geography, groups, and in the treatment of people and distribution of income and wealth should give us good opportunity for rethinking of the problem. Mitigating measures being implemented during the time of Coronavirus pandemic have highlighted misgivings about current narratives on economic benefits and accruals, in both developed and under-developed world. Underlying shortcomings have also challenged prevailing *laissez-faire-'ism'* and its unfettered relevance that had divided elites, scholars, academics, practitioners, planners, and policy makers for decades in making choices between '***people's life***' and '***sustenance livelihood***.'' Energy and resources largely squandered debating to ensure a balance and differences demonstrates our collective inability to discover the right political synergies for keeping social harmony and economic cohesion. And, present situation is even further stressed by the unexpected force generated by the power of social media for it not only produces action, reactions and counteractions; it produces enormous confusion by creating strong over-reactions at large scale. This is unprecedented in evolution of human society and economic history. We are, thus, confronted with a complex set of new challenges that global leadership must commit to enhancing sustainable development measures for eradicating economic inequalities because inequality hurts human growth and dampens inclusive prosperity.

In this complex, climate emergency, the most alarming challenge of 21st Century, if combined with global pandemic one ought to invent new economic thought of the day, which can work across all communities and societies and help improve universal prosperity, wellbeing, and improve quality of life. This means being sensitive to social values, ethics, economy, environment, and reduce excessive consumption that is depleting natural resources beyond replenishing, and not just work for glory, and serve privileged and wealthy. It is now time to put on-hold all excesses and insatiable thrust for infinite greed and dominance.

This book will provide multi-level understanding of multiple perspectives on reasons, phenomenon of economic inequality and its fallouts.

Professor Medani Bhandari has done it again and produced this volume highlighting core problems. This publication is timely. It will enlighten the world community on challenges facing economic inequalities at a time when United Nations Sustainable Development Goals targets, adopted prior to pandemic, are under stress for most of the goals might not be met by the target year 2030 if international efforts are not re-doubled in post-pandemic gloomy world of socio-political upheavals and uncertain economic environment.

Kedar Neupane is Executive Board Member of Nepal Policy Institute, and retired staff of United Nations High Commissioner for Refugees. (Views expressed in this are his personal opinion.) Feb. 3, 2021.

Endorsements

The book's title is catchy and invokes some curiosity as to what conclusions this book brings to the readers in the backdrop of so many books and literatures written on the subject of economic inequality. The list of topics to be included in the book certainly arouse my interest. Have I had the benefit of reading some chapters of this book, that would certainly help me make some relevant comments that would do the justice to the book. Economic inequality is a heavy subject that, I believe, falls under the domain of political economy. Any scholarly attempt to examine the complex issues of political economy is certainly a welcome initiative. I believe, the book will be a useful reference to policymakers, educators, development planners, and practitioners.

Prof. Gopi Upreti, Fairfax, Virginia, USA Date: 03/22/2021

Social inequalities in terms of gender, age, class, color, ethnicity, religion, and disability persist globally. Despite the tremendous efforts and progress made, large disparities and discriminations remain worldwide in terms of access to food, health, education, judiciary services and right to vote, and especially in least developed and developing countries. Such inequalities and discriminations have resulted in cultural, social, and economic unrest and conflicts in society and widening gap between haves and have-nots. This is a comprehensive book dealing with all aspects of social inequalities. Prof. Medani P. Bhandari and Dr. Hanna Shvindina have done an excellent job of authoring and editing this book that includes chapters by eminent authors from several countries highlighting the extent, problems, and consequences of social inequalities. This book attempts to close the social inequality gap in the society and achieve the SDGs related to social inequality, especially SDG 10. I found it rich in facts and easy to read and understand different aspects of social inequalities prevailing and is a helpful resource for ideas on how social inequality can be reduced in the society. I endorse the book as there are hardly such books available in the market that capture the depth of information that this book has provided.

Dr Jagadish Timsina, Editor-Agricultural Systems, Melbourne, Australia

Inequality has become a major challenge facing the world. Its causes are simultaneously political, economic, and social. The escalating level of inequality has led to social instability, crisis, and conflicts around the world. While national governments and United Nations recognize this as a major challenge, too little is being done on the ground to overcome this. This book, edited by Prof. Medani P. Bhandari and Dr. Shvindina Hanna, offers a timely and critical analysis of socio-economic foundations of inequality, covering all key parameters such as income and wealth distribution, equity & equality of outcome, and equality of opportunities. With chapters written by eminent scholars using latest research and incisive analysis, the book is an important primer for understanding the nuances of economic inequality in the contemporary world.

Dr. Hemant Ojha, Associate professor, University of Canberra, Australia.

Acknowledgements

First of all, I would like to thank my colleague, co-editor of this volume Associate Professor Shvindina Hanna and all members of her family (parents, Olexandr and Valentina) and Prof. Volodymyr Boronos, Prof. Tetyana Vasilyeva, Prof. Oleg Balatskyi (her mentor 1999–2012). I should acknowledge that, she is the key manager of this project—making an initial book plan to establishing contacts with authors—collecting manuscripts, and other editorial tasks as needed.

Equally, I would also like to thank all authors who contributed to this volume—Drs./ Profs. *Man Bahadur Bk; Sergii Polumiienko; Liubov Zharova and Natalia Eremeeva; Tetyana Vasilyeva, Serhiy Lyeonov, Olena Pakhnenko and Pavlo Rubanov; Carlos Sanchez; Jacob, D. E., and I. N. Ufot; Ievgen Khlobystov and Lidiia Horoshkova; Yuliia Matvieieva, Yuliia Opanasyk, Olena Pavlenko and Iuliia Myroshnychenko; Yuriy Petrushenko, Anna Vorontsova and Oksana Ponomarenko; Anna Rosohata and Liubov Syhyda; Olha Kuzmenko, Anton Boyko and Victoria Bozhenko; Olena Chygryn, Tetyana Pimonenko, Oleksii Lyulyov,* Each of you have done an excellent job, your scholarly work bridges the knowledge gap in the complex field—inequality with reference to sustainable development goals.

I would like to remember Prof. Bishnu Paudel (who passed away due to heart attract in early 2020). Prof. Poudel was the motivator, encourager, and supporter for me to continue my collaborative work to address the inequality challenges around the world. His understanding of Bashudhiva Kutumbakam (all living being are our relatives and neighbors) was slightly different than ordinary meaning. He used to say that the essence of Bashudhiva Kutumbakam is to see everyone as you see yourself, love everyone as you love yourself. He always used to say that any contribution toward addressing the challenges like inequalities, climate change, sustainability, pandemic, and other social, economic, political, and natural challenges are the pathway to change human behavior toward Bashudhiva Kutumbakam (all living being are our relatives and neighbors).

I would also like to thank Prof. Douglas Capogrossi (my mentor, Emeritus, President of Akamai University, Hawaii), Dr. Mary Jo Bulbrook,

(President, Akamai University), Dr. Ambika Adhikari (mentor for environment conservation), Dr. Krishna Prasad Oli (Member of National Planning Commission, Government of Nepal); Prof. Keshav Bhattarai (Arizona State University), Mr. Kedar Neupane (former staff of UNHCR), Prof. Durga D. Poudel, University of Louisiana at Lafayette, Lafayette, Louisiana, USA, Dr. Jacek Piotr Binda (Poland), Dr Jagadish Timsina, and Dr. Hemant Ojha, (Australia), for their reviews, forwards, and endorsements. Your togetherness with us adds value in this book project.

Special thanks to Prof. Boronos Volodymyr, Head of the Department, of Finance and Entrepreneurship, I will never forget your encouragements during my stay in Ukraine. You have given me new insights especially to dig into Marxist philosophy. I would also like to thank all colleagues of Department of Finance and Entrepreneurship. You all are amazing, your knowledge sharing expertise is commendable. Thank you to you all.

I would also like to thank my wife Prajita Bhandari for her encouragement and for giving insightful information on how women are victims of inequality—at home to work environment. Without your support, this project was impossible. I would also like to thank our son Prameya, daughter Manaslu and daughter-in-law Kelsey for helping me find relevant resources in the field of inequality, social division, stratifications, and gender issues. I would also like to thank our granddaughter for giving us a joyful environment, which always helps to concentrate in the work I am tuned in. I would also like to thank our son-in-law Abhimanyu Iyer, and his parents Mahesh and Uma Iyer for insightful thoughts on inequality issues. I would also like to thank my mother Hema Devi, brothers Krishna, Hari, sisters Kali, Bhakti, Radha, Bindu, Sita and their families for encouraging me by providing a peaceful environment. I would also like to thank my friends Kshitij Prasai, Dr. Prem Bhandari, Dr. Man Bahadur Khari, Rajan Adhikari, Govinda Luitel, Medini Adhikari, Prof. Gopi Uprety, Tirtha Koirala, Prof. Sanjay Mishra, Dr. Bijay Kattel (my mentor for wildlife conservation in Nepal), Dr. Aleksander Sapinski (Poland), Dhir Prasad Bhandari and all my Facebook and LinkedIn's friends, who have been always encouraging to us to give back to the society through knowledge sharing. Thank you to Rajeev Prasad, Junko, and all friends of Rivers Publisher, for encouraging and empowering us to complete this book project on time. Thank you to you all, who have given their inputs for this book project directly or indirectly.

Thank you,

Medani P. Bhandari

List of Figures

List of Tables

1

Trends, Traps and Trade-offs of Economic Inequality—A Growing Pattern and Concern

Medani P. Bhandari[1] and Shvindina Hanna

[1]Prof. of Sustainability, Akamai University, Hilo, Hawaii, USA medani.bhandari@gmail.com

1.1 Introduction

Inequality is a chronic divisive factor of society. It is a known truth/fact that several inequalities (economic, social, cultural, religious, geographical, etc.) have been omnipresent in human society. Inequalities can be found within each family, from family to community, from communities to the nation, and globally. Inequality is the major cause of political, economic, and social instability that creates crisis and conflict within the society and nation, and then in other nations. The major causes of inequality are unequal, uneven, biased, and power-centric distributions of human, economic, social, political, cultural, and spiritual necessities. Inequality in one sphere of system has interwoven relationships with all social systems (including economic). Among all, economic inequality is the one, which dominates the entire social system; therefore, economic inequality is the most discussed issue in the academia as well as in the economic development discourses. "*Many authors distinguish "economic inequality", mostly meaning "income inequality", "monetary inequality" or, more broadly, inequality in "living conditions""* (United Nation 2018:1).

As stated above, most of the discussions about inequality are measured through economic lenses, primarily through development economics. Development economics captures the dimensions of inequality, with the focus of well-being of the poor nations. However, socially and culturally grounded circumstances, ethnicity, family background, gender, sexual orientation, etc. as well as human's talent and effort are divisive factors that are not captured in the development paradigms (Bhandari and Shvindina 2019).

Economic inequality is considered as one of the major challenges. Many international organizations, governments, non-governmental organizations, and other agencies are working to reduce inequality. United Nations (2015) clearly acknowledges that inequality is one of the major challenges of the contemporary world. As an evidence "Out of the 17 goals, eleven address forms of inequality, in terms of equality, equity, and/or inclusion (Goals 1, 3, 4, 5, 6, 7, 8, 9, 10, 11, 16 and 17), and one goal (Goal 10) explicitly proposes to reduce various forms of inequalities" (Freistein and Mahlert 2015:7). Chapters of this book directly or indirectly reveals how each goal are interconnected and point out the difficulties of achieving desired goals and targets of sustainable development listed in the United Nation 2015 declaration "Transforming our world: the 2030 Agenda for Sustainable Development" (Bhandari and Shvindina 2019).

However, despite various efforts, economic inequality is still increasing. The applied measures to reduce inequality by the international organizations, governments, and other concerned stakeholders seem to be not sufficient or inadequate. Chapters of this book have tried to critically examine the causes and consequences of economic inequality.

To minimize economic inequality, it is necessary to break the webs created by the traditionally inherited elites as well as the newly emerged haves group, and empower, encourage, insist, and uplift the have-nots group, in a way that the dominated, marginalized group will also get equal opportunities to survive, grow, and enjoy their lives.

1.2 What Is the Future of Global Income Inequality?

"The future of global income inequality is likely to be shaped by both convergence forces (rapid growth in emerging countries) and divergence forces (rising inequality within countries). No one knows which of these forces will dominate and whether these evolutions are sustainable.

However, our benchmark projections show that if within-country inequality continues to rise as it has since 1980, then global income inequality will rise steeply, even under fairly optimistic assumptions regarding growth in emerging countries. The global top 1% income share could increase from nearly 20% today to more than 24% in 2050, while the global bottom 50% share would fall from 10% to less than 9%.

If all countries were to follow the high inequality growth trajectory followed by the United States since 1980, the global top 1% income share would rise even more, to around 28% by 2050. This rise would largely be made at the expense of the global bottom 50%, whose income share would fall to 6%.

Conversely, if all countries were to follow the relatively low inequality growth trajectory followed by Europe since 1980, the global top 1% income share would decrease to 19% by 2050, while the bottom 50% income share would increase to 13%.

Differences between high and low inequality growth trajectories within countries have an enormous impact on incomes of the bottom half of the global population. Under the US-style, high inequality growth scenario, the bottom half of the world population earns €4 500 per adult per year in 2050, versus €9 100 in the EU-style, low inequality growth scenario (for a given global average income per adult of €35 500 in 2050 in both scenarios)" (World Inequality Report 2018:250).

1.2.1 How inequality can be minimized?

A capability is a potential functioning; the list of functioning is endless. It might include doings and beings such as being well nourished, having shelter and access to clean water, being mobile, being well-educated, having paid work, being safe, being respected, taking part in discussions with your peers, and so on. The difference between a capability and functioning is like one between an opportunity to achieve and the actual achievement, between potential and outcome (Walker 2006: 165).

> *"We must all realize that inequality reduction does not occur by decree; neither does it automatically arise through economic growth, nor through policies that equalize incomes downward via blind taxing and spending. Inequality reduction involves a collaborative effort that must motivate all concerned parties, one that constitutes a genuine political and social innovation, and one that often runs counter to prevailing political and economic forces" (*Genevey *et.al. 2013:6).*

However, in practice, there is no seriousness to overcome the actual challenges of inequalities embedded within the socio-political, economic, cultural, traditional, and religious practices. There are some improvements in social, political, cultural, traditional, and religious inequalities in the developing world; however, they are not being able to address or reduce the economic inequality. The factual truth is that the economic inequality has increased in both developed and developing countries (Bhandari and Shvindina 2019). The gap between rich and poor is increasing. Even due to the COVID-19, this gap has widened (more research is needed to present the increased actual economic inequality gap within a year or so). However, it is true that, the world lost thousands of people, due to COVID and thousands of people have lost their jobs, facing starvation, insecurity, and facing the challenge for survival.

The impact of COVID-19 has several folds—short term and long term. We can assume that the impact of COVID-19 will remain for several years and the wound inflicted by it will never be healed completely. The challenges of inequality could be more severe, and the gap of haves and have-nots could increase in alarming rate in the coming years. Therefore, it is extremely important that, all concerned stakeholders (including governments, academic scholars, freelance thinkers, international organizations (the United Nations, World Bank, IMF), Regional Development Banks, multinational agencies, international non-governmental organizations, national non-profit and profit organizations, industries, etc.) need to be prepared to overcome the consequences of COVID-19 and variation between haves and have-nots people of the world. This book's chapters show how economic inequality has divided the world and how and why rich are getting richer and how poor are becoming poorer unevenly, particularly in the developing world.

1.3 The Economic and Social Burden is Uneven

In the developing world, the economic and social burden is uneven between males and females, between the rich and the poor, and between the socially dominant group and the dominated group. Females have been always the victims of the traditionally bounded society. As shown in the cover picture of the book, maintaining the house is always a woman's responsibility.

Picture 1—Route Tribe of Nepal: source: Dr. Man Bahadur BK (one of the chapter contributor).

This picture is from Western Nepal, taken in December 2020; where a woman is carrying six pots in her head depicts the level of burdens the young women face. There are 8 people in this picture, and each of them is watching her in a way that she is carrying the hope, the medicine of thirst and hunger. In the background, on the right side, there are two adult women and a walking baby; on the left, an adult woman and a baby; on the right side, an old man sitting on the bare stony surface; and on the right front, a baby girl just walking behind her. The picture also shows the huts (houses) where they live. Needless to state that, this picture not only shows how the life of the poor in the developing countries is but also reveals the factual truth of economic and social inequality.

Picture 2—Route Tribe of Nepal: source: Dr. Man Bahadur BK

As shown in the first picture, second picture gives a closer view of the hardships of the poor and how they are surviving. On the back, two women are standing with sticks on their hands, babies are on the rocky floor; in the middle, an old man is preparing firewood, and beside him a woman is sitting with her baby. In the front, a baby girl is sitting on the bare rocky ground and making a leaf doll. The huts are their homes. Can these huts hold the rain, provide actual shelter?

Picture 3—The blacksmith: source: Dr. Man Bahadur BK

The exercise of survival—Western Nepal: A poor blacksmith is in his traditional homemade woven, where they makes knifes etc. He relies on the income that he generates through his traditional occupation. His occupation is not safe, because people can buy cheap knifes made in modern factory and his occupation is in danger. What is the impact of mechanization on rural livelihood pattern?

Picture 4—Traditional Occupation is in Danger—Eastern Nepal: source: Dr. Man Bahadur BK

The young man in the picture is making baskets for carrying heavy goods, mostly used by the porters. However, the demand for such strong bamboo basket is diminishing due to the growing use of western made heavy backpacks (bags).

Picture 5—Hunger is hunger and survival is the first priority.
Source: Dr. Man Bahadur BK.

These three women and five girls are enjoying their food, needless to say where and in which condition. We can see the garbage beside them. Is this a healthy place for food? Why are they eating there? Why all women and girls?

These pictures above provide the reality of inequality, reality of social and economic burdens on women and men who are the victims of the divisive world.

> A girl born to a lower-caste family of nine in the slums of Dhaka has vastly different opportunities from a boy born to well-educated and affluent parents in the well-heeled neighborhoods. An AIDS orphan in rural Zimbabwe is almost certain to have fewer chances and choices in life than a compatriot born to healthy and well-educated parents in Harare. Those differences are even greater across borders: an average Swiss, American, or Japanese child born at the same instant as one in a poor, rural area of South Africa will have incomparably superior life chances.
>
> Such staggering inequalities in opportunity are intrinsically objectionable, and almost every culture, religion, and philosophical tradition

> has developed arguments and beliefs that place great value on equity for its own sake. But that is not all. We would argue—as elaborated in the World Bank's World Development Report 2006: Equity and Development—that there is now considerable evidence that equity is also instrumental to the pursuit of long-term prosperity for society as a whole. …The shaping of opportunities begins before individuals are born. Who one's parents are, what country they live in, and how rich they are, make a large difference to a person's opportunities in terms of life expectancy, education, access to services, and economic prospects. … Infant mortality rates vary markedly both within and across countries. In El Salvador, for instance, babies born to mothers with no schooling are four times more likely to die before their first birthday than babies whose mothers are better educated (Ferreira, Francisco H.G. and Michael Walton 2005).

Economic inequality problems were analyzed in different aspects by previous researchers, the trade-offs between economic growth and instability are in the focus of many researchers. The economic growth does not guarantee the economic stability, and stability does not mean economic growth neither. The inequality rises as the consequence of economic development and resources reallocation, at the same time without the differences between regions, nations, and individuals there will be no dynamic. Among all the regions worldwide Ethiopia, Zambia, South Africa, Kenya, and South Sudan are TOP-5 countries with the highest interest to the issues of economic inequality according to the recent trends (see Figure 1.1) in Google search.

The inequality trends are observed all around the world, and since 1980 there is an income inequality dynamic shaped by a diversity of factors, including environmental, institutional, political, and technological factors. The income and, therefore, economic inequalities can be linked with educational opportunities as well.

Inequality trap(s) is a quite new term that describes the inequalities in opportunity reproduced over time and across generations, through economic, sociocultural, and political mechanisms (Ferreira & Walton, 2005). In other words, it is a vicious circle of inability to get access to the opportunities for those who are most deprived of it.

Thus, the trade-offs, trends, and traps brought by inequality became the main subject of the book. On the basis of few conceptual notion and the practical challenges of inequality, this edited book "Economic Inequality—Trends, Traps and Trade-offs" presents the unexplored issues of economic

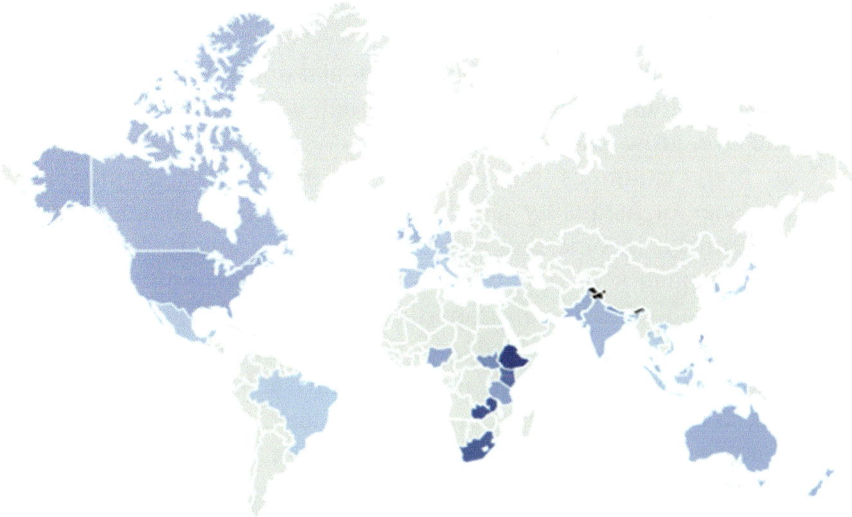

Figure 1.1 Interest by region to "economic inequality."
Source: designed via Google Trends for the base keyword.

inequality, with the exemplary case studies of various countries. Inequality is a chronic divisive factor of society. It is a known truth/fact that various inequalities (economic, social, cultural, religious, geographical, etc.) have been omnipresent in human society. Inequalities can be found within each family, from family to community, and from communities to the nation, and globally. Inequality is the major cause of political, economic, and social instability that creates crisis and conflict within the society and nation, and then in other nations. The major causes of inequality are unequal, uneven, biased, and power-centric distributions of human, economic, social, political, cultural, and spiritual necessities. The human civilizations are grounded in a way that, small elite groups create webs and networks within which they lead the society, rule the society in a way that leads public rarely became capable enough to get opportunities even to participate in fair competitions. In marginalized societies, individuals would not have similar access to wealth, education, health, and other benefits as the ruler elites have. Therefore, the elite groups remain elite for many generations, and the dominated groups remain as victims. These scenarios can be seen as political inequality, which is the major cause of inequality in the society. The dominant group capture the political opportunities and power of governance so that marginalized group always remain at the lower strata of the political, social, and economic system. To minimize economic inequality, it is necessary to break the webs created by the traditionally inherited elites as well as from the newly emerged haves group; and empower, encourage, insist, and uplift the have-nots group, in a way that the dominated, marginalized group will also get equal opportunities to survive, grow, and enjoy their lives.

This book examines the major parameters of intertwined socio-economic issues of inequality and focuses on the key economic issues of inequality, namely, income and wealth distribution, equity & equality of outcome, and equality of opportunities. Basically, economic inequality is measured by wealth, income disproportions in distribution and consumption patterns in the specific area. Mostly, for the statistical presentation, inequality is measured using Gini Coefficient, inequality adjusted human development index, 20:20 ratio, Palma ratio, Hoover index, Galt score, Coefficient of variation, Theil index, wage share, etc. However, it is not necessary that all income can be measured in the monetary unit and can be presented and proved through statistical tools. In this book, authors tried to unveil the problems and consequences of socio-economic inequality through case studies of different countries. Based on various case studies, this book encourages to rethink economic development through the lens of growing inequalities and disparities. These disparities became obvious with sharing information about global changes and distribution of environmental, economic, and social risks among the recipients. The economic growth per se is disproportional, and the efforts of scholars, practitioners, and policymakers should be directed to empower the marginalized society in a way that no one should get left behind (UN Slogan). The main challenges and focus of every chapter are different, and yet in combination, they give an integrated understanding of the phenomenon of economic aspects of inequality.

References

1. Berg, Andrew, and Jonathan Ostry (2011). Inequality and unsustainable growth: two sides of the same coin? IMF Staff Discussion Note. SDN/11/08. Washington, D.C.: International Monetary Fund. 8 April.
2. Bertelsmann Stiftung and Sustainable Development Solutions Network (2018), SDG Index and Dashboards Report 2018-Global Responsibilities, Implementing the Goals, G20 and Large Countries Edition. www.pica-publishing.com, http://www.sdgindex.org/assets/files/2018/00%20SDGS%20 2018%20G20%20EDITION%20WEB%20V7%20180718.pdf
3. Bhandari, Medani P. and Shvindina Hanna (2019), The Problems and consequences of Sustainable Development Goals, in Bhandari,

Medani P. and Shvindina Hanna (edits) Reducing Inequalities Towards Sustainable Development Goals: Multilevel Approach, River Publishers, Denmark / the Netherlands- ISBN: Print: 978-87-7022-126-9 E-book: 978-87-7022-125-2

4. Ferreira, Francisco H.G. and Michael Walton (2005), The Inequality Trap, IMF Quarterly Bulletin, December 2005, Volume 42, Number 4, https://www.imf.org/external/pubs/ft/fandd/2005/12/ferreira.htm
5. Fobes (2019), America's Wealth Inequality Is At Roaring Twenties Levels, Contributor-Jesse Colombo, Forbes (https://www.forbes.com/sites/jessecolombo/2019/02/28/americas-wealth-inequality-is-at-roaring-twenties-levels/#62f244642a9c).
6. Freistein, K., and. Mahlert, B. (2015), The Role of Inequality in the Sustainable Development Goals, Conference Paper, University of Duisburg-Essen
   See discussions, stats, and author profiles for this publication at: https://www.researchgate.net/publication/301675130
7. Genevey R. (ed.), Pachauri R.K. (ed.), Tubiana L. (ed.), Jozan R., Voituriez T., Sundar S.. (2013). A planet for life : sustainable development in action, TERI, New Delhi http://regardssurlaterre.com/sites/default/files/edition/2016/01-PFL%202013%20Inequalities.pdf
8. Guidetti, Rehbein (2014), Theoretical Approaches to Inequality in Economics and Sociology, Transcience, Vol. 5, Issue 1 ISSN 2191-1150 https://www2.hu-berlin.de/transcience/Vol5_No1_2014_1_15.pdf
9. Jan-Peter Voß, Jens Newig, Britta Kastens, Jochen Monstadt† & Benjamin No¨ Lting (2007), Steering for Sustainable Development: a Typology of Problems and Strategies with respect to Ambivalence, Uncertainty and Distributed Power, Journal of Environmental Policy & Planning Vol. 9, Nos. 3-4, September–December 2007, 193–212 https://www.researchgate.net/profile/Jochen_Monstadt/publication/233049753_Steering_for_Sustainable_Development_A_Typology_of_Problems_and_Strategies_with_Respect_to_Ambivalence_Uncertainty_and_Distributed_Power/links/577ff29608ae5f367d370a97/Steering-for-Sustainable-Development-A-Typology-of-Problems-and-Strategies-with-Respect-to-Ambivalence-Uncertainty-and-Distributed-Power.pdf
10. United Nations (2020), World Social Report 2020, Inequality in A Rapidly Changing World, Department of Economic and Social Affairs, New York https://www.un.org/development/desa/dspd/wp-content/uploads/sites/22/2020/02/World-Social-Report2020-FullReport.pdf https://www.un.org/development/desa/dspd/wp-content/uploads/sites/22/2020/01/WSR2020-Executive-Summary.pdf

11. United Nations (2020a), Inequality – Bridging the Divide- Shaping our future together- United Nations 75-2020 and Beyond, United Nations, New York https://www.un.org/en/un75/inequality-bridging-divide
12. Walker, M. (2006) Towards a capability-based theory of social justice for education policymaking. Journal of Education Policy, 21(2), 163-185.
13. *World Inequality Report (2018), The World Inequality Lab,* https://wir2018.wid.world/ https://wir2018.wid.world/files/download/wir2018-summary-english.pdf

2

Liberal Socialism: Path Forward to Reset the CORONOMICS[1]

Dr. Man Bahadur Bk

Former Principal Secretary, State Government of Bagamati, Nepal

ABSTRACT

The global economy has been slowed-down due to the COVID-19 pandemic almost throughout the year. Most of the countries have imposed containment as a precautionary measure. Most international flights have been shut down and production and trading businesses have been no more in operation. The human-being came back to its humane character from its robotized behavior. Till November 2, the death toll crossed 1.2 million predominantly in developed countries led by the USA. Although the virus transmission is worldwide, the severity of its outbreak is relatively higher in developed countries. It has reversed the world economy creating the greatest recession around the globe since 1930s. The pandemic that the world has been suffering is the fundamental one, but there are multiple devastating effects on human lives and livelihoods tending toward a severe hunger epidemic. Through the experiences, it is justified that the neoliberal economic policy does not have adequate resilience. Capitalism has generated massive wealth for some, but it has devastated the planet and has failed to improve human well-being. So, the emergence of a new policy that balances economic mobility and human well-being is sought. The concept of "Protective Liberalism" can be a replacement to libertarianism as the socio-economic policy for "Liberal Socialism." It can also be coined as "Social Capitalism" that resets "Coronomics" in an egalitarian manner.

Keywords: Protective Liberalism, Liberal Socialism, Coronomics, Social Capitalism.

[1] *CORONOMICS is the term coined for the economic turmoil due to the COVID-19 pandemic.*

2.1 Background

The global economy has been heavily slowed-down due to the COVID-19 pandemic almost throughout the year. Some countries have formally imposed the lockdown and in some countries, people keep themselves in self-isolation, a precautionary measure as the threat of the virus transmission grows higher. Most international flights remained shut down and production and trading businesses have not been in operation for a significant period. It seems everything remained in long rest; the earth has enjoyed its ancient nature recharging the fullest. The human-being came back to its humane character from the robotized-mechanical behavior.

Till November 2, the death toll reached 1,206,810[2] predominantly in developed countries led by the USA. The amazing fact is that even though it evolved in China, the spread jumped to the Europe and America at its initial phase. Figure 2.1 presents the rise in the wave of the COVID-19 transmission. Although the virus transmission now is worldwide, the initial severity of its outbreak was relatively higher in countries such as Europe and the US, which favor and practice the liberal economy, while comparatively countries with less liberal (closed) economy such as Cuba, North Korea, some Russian and South Asian countries were less affected.[3] It shows that the open economy in the name of neoliberalism loses its resilience to control pandemic. With COVID-19 outbreak, we are not only facing a global health pandemic but also a global humanitarian catastrophe.[4] Moreover, it has reversed the

Table 2.1 Death toll due to COVID-19 until Nov 2, 2020

Region	Population (2020)	Total Cases	Total Deaths	Total Recovered
World	7,794,798,739	46,936,520	1,206,810	33,812,533
Europe	747,636,026	10,266,615	268,330	3,875,330
North America	368,869,647	11,320,887	353,604	7,480,529
Asia	4,641,054,775	13,794,436	245,085	12,278,114
South America	653,962,331	9,709,533	295,535	8,672,213
Africa	1,340,598,147	1,807,556	43,286	1,473,659
Oceania	42,677,813	37,493	970	32,688

Source: https://www.worldometers.info/ visited on Nov. 2, 2020.

2 https://www.worldometers.info/ visited on November 2, 2020

3 https://www.academia.edu/42328909/Revival_of_Global_Economy_after_COVID-19_Pandemic

4 https://www.wfp.org/news/wfp-chief-warns-hunger-pandemic-covid-19-spreads-statement-un-security-council

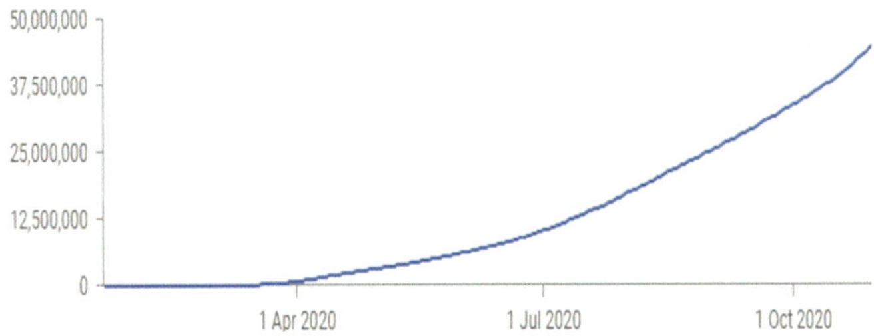

Figure 2.1 Total cases worldwide over time.
Source: https://news.google.com/covid19/map

world economy creating the greatest recession around the globe since 1930s. However, it offers a rare chance for the global community to undertake several policy changes to address the public health challenge and reset the global economy through preventive and protective measure. So, the economist Joseph Schumpeter called it the "gale of creative destruction" that refers exclusively to creative economic transformation.[5]

2.2 Impact of the Pandemic

After a decade of uninterrupted growth, the global economy came to a sudden halt because of the COVID-19 pandemic.[6] The global recession has already been started with this pandemic but the question is how deep and how long it will be. Even before the COVID-19 outbreak, most developing countries were already on a shakier economic footing than they were in the run-up to the 2009 global recession.[7] So, the developing economies likely to be hit the hardest—those dependent on trade, commodities or tourism—tending to have higher concentrations of people in extreme poverty. The pandemic that the world has been suffering is the fundamental issue but there are multiple devastating effects on human lives and livelihoods due to the economic stagnation. The foremost impact is on the job loss and disruption in distribution chain since the production and market are no more functioning.[8] People are

---

[5] *https://blogs.adb.org/blog/how-covid-19-crisis-can-spur-economic-transformation*

[6] https://blogs.worldbank.org/voices/scaling-covid-19-crisis-response-now-will-avoid-higher-costs-later

[7] Ibid.

[8] Bhandari, Medani P. (2020), The Phobia Corona (COVID 19) - What Can We Do, Scientific Journal of Bielsko-Biala School of Finance and Law, ASEJ 2020, 24 (1): 1-3, GICID: 01.3001.0014.0769, https://asej.eu/resources/html/article/details?id=202946

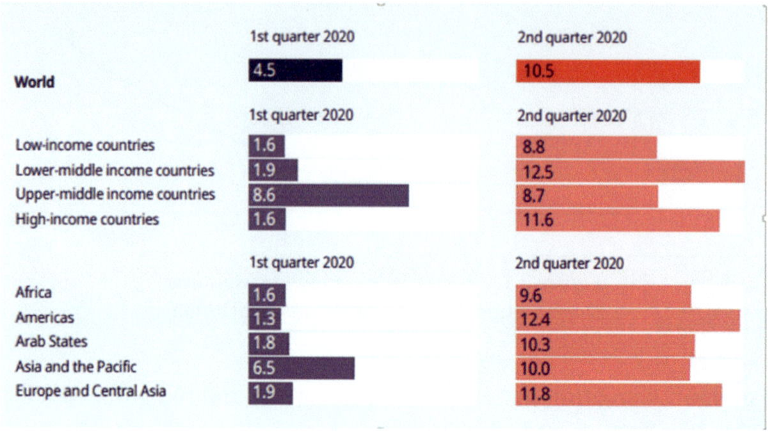

Figure 2.2 Estimated drop in aggregate working hours, globally, by region and by income group (%).
Source: ILO, 2020

losing their resilience since the jobs and other means of livelihoods have been halted. The ILO assessment reflects the full or partial lockdown measures affecting almost 2.7 billion workers—four in five of the world's workforce.[9] Losing job leads to further losing affordability for the food stuffs and other daily consumables. The latest ILO data on the labor market impact of the COVID-19 pandemic reveals a 10.5 percent drop in working hours, equivalent to 305 million full-time jobs (assuming a 48-hour working week) in the second quarter of the year due to the prolongation and extension of lockdown measures (ILO, 2020). There is the devastating effect on workers in the informal economy and on hundreds of millions of enterprises worldwide. Among the most vulnerable in the labor market, almost 1.6 billion informal economy workers (out of a worldwide total of two billion and a global workforce of 3.3 billion) have suffered massive damage to their capacity to earn a living.[10] The first month of the crisis is estimated to have resulted in a drop of 60 percent in the income of informal workers globally. This translates into a drop of 81 percent in Africa and the Americas, 21.6 percent in Asia and the Pacific, and 70 percent in Europe and Central Asia. Workers in four sectors that have experienced the most "drastic" effects of the disease and falling production are food and accommodation (144 million workers), retail and wholesale

[9] https://news.un.org/en/story/2020/04/1061322
[10] https://www.ilo.org/global/about-the-ilo/newsroom/news/WCMS_743036/lang--en/index.htm

(482 million), business services and administration (152 million), and manufacturing (463 million) (Lal 2020). Without alternative income sources, these workers and their families will have no means to survive. Worldwide, more than 436 million enterprises face high risks of serious disruption. These enterprises are operating in the hardest-hit economic sectors. In developing countries, risks of food insecurity have been emerged. Some 821 million people used to go to bed hungry every night all over the world even before the pandemic, and the latest Global Report on Food Crisis reveals that 135 million people on earth are marching toward the brink of starvation. Moreover, the WFP analysis shows that, due to the current pandemic, an additional 130 million people could be pushed to the brink of starvation, which is a total of 265 million people, by the end of 2020.[11] The WFP estimates that now 1.6 billion children and young people are currently out of school due to lockdown closures. Nearly 370 million children are missing out on nutritious school meals.[12] There is the proximity of famine in the least developed countries due to the prolonged lockdown and distribution halt. Thus, the health pandemic is tending toward a severe hunger epidemic.

Lockdowns and economic recession are expected to lead to a major loss of income among the working poor. It will have hard-hit even harder to the hardcore poor like daily wage-earners, slum-dwellers, disabled people, elders, refugees and migrant workers. Massive job loss will also be hard-hit to the labor-sending countries since there will be heavy fall in remittances. Global remittances peaked at $706 billion in 2019, this safety net is now frayed by job losses in the service sectors most reliant on migrant workers.[13] The WB estimates a reduction of 20 percent of the remittances in 2020. Even small changes in remittance policy can have a big effect on financial systems in developing countries over time. Likewise, the loss of tourism receipts will damage countries' economy which are significantly based on tourism industry. The collapsing oil prices in lower-income countries like South Sudan will have an impact significantly, where oil accounts for 98.8 percent of total exports.[14] The 700 million, close to 10 percent of the world population continues to live below the extreme poverty line of $1.90 per day and over half of the world's extreme poor live in middle-income countries, with

[11] https://www.wfp.org/news/wfp-chief-warns-hunger-pandemic-covid-19-spreads-statement-un-security-council

[12] https://www.wfp.org/news/wfp-chief-warns-hunger-pandemic-covid-19-spreads-statement-un-security-council

[13] https://blogs.worldbank.org/psd/remittances-times-coronavirus-keep-them-flowing

[14] https://blogs.worldbank.org/psd/remittances-times-coronavirus-keep-them-flowing

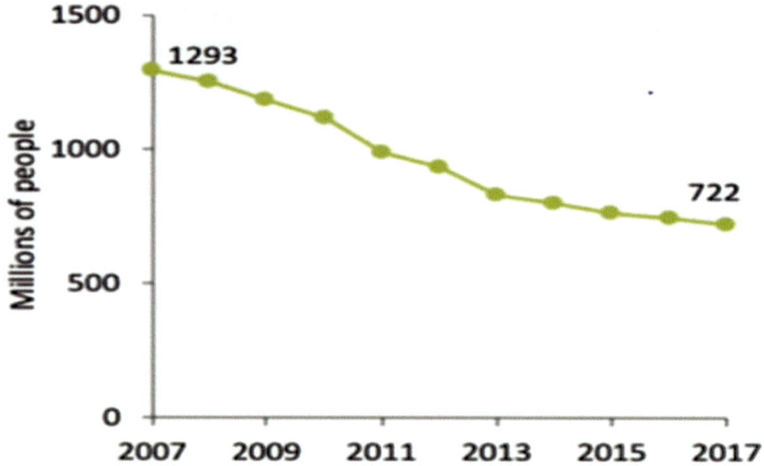

Figure 2.3 Millions of people living in extreme poverty, 2007–2017.
Source: The World Bank, 2020

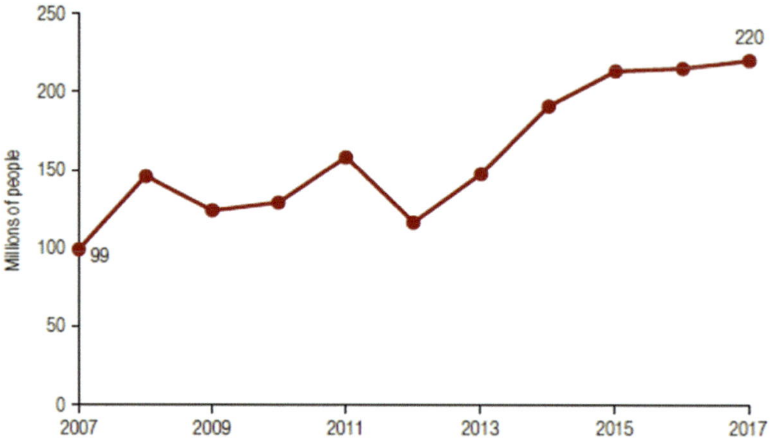

Figure 2.4 The World's population living in proximity to conflicts, 2007–2017.
Source: The World Bank, 2020

India and Nigeria together accounting for roughly one-third of the extreme poor (WESP 2020). Still the 1.3 billion people are without access to basic needs (UN 2020). The WB estimates close to 90 million people expected to fall into extreme deprivation this year.[15] Between present disruptions and

[15] *https://blogs.imf.org/2020/10/13/a-long-uneven-and-uncertain-ascent/*

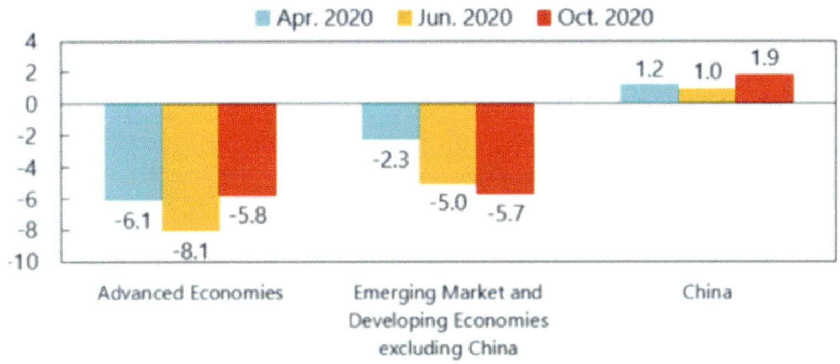

Figure 2.5 Uneven recoveries.
Source: IMF, 2020

future threats to the food supply chain, the pandemic has generated extreme vulnerability in the agriculture sector. Water insecurity is set to hit the most vulnerable the hardest as the 40 percent of the world population lack access to water and sanitation.[16] It is the fact that the world's success is driving down extreme poverty is overshadowed by failure to reduce fragility, conflict, and violence. Still today more than two billion people are living in proximity to conflict, increased from 1.5 percent in 2007 to 3.0 percent in 2017 (WB 2020). At the same time, the number of forcibly displaced people worldwide has more than doubled, exceeding 70 million in 2017. The present pandemic will certainly rise the number of these people creating more vulnerability for life and livelihood.

According to IMF October report (2020) the world economy grew at 2.8 percent in 2019. However, for the first time since the Great Depression in the 1930s, both advanced economies and emerging market and developing economies are in recession. As a result of the pandemic, it is projected to contract sharply by –4.4 percent in 2020, much worse than during the 2008–09 financial crisis. In a baseline scenario, which assumes that the pandemic fades in the second half of 2020 and containment efforts can be gradually unwound, the global economy is projected to grow by 5.2 percent in 2021 as economic activity normalizes, helped by policy support. Growth in the advanced economy group, where several economies are experiencing widespread outbreaks and deploying containment measures, is projected at –5.8 percent in 2020 and expected to grow by 3.9 percent in 2021. Likewise, the

[16] https://www.academia.edu/42686038/Impact_of_COVID-19_Pandemic_on_Food_Security_Escalating_World_Hunger

Table 2.2 Overview of the World Economic Outlook projections (in percentage)

		Projections	
Year	**2019**	**2020**	**2021**
World Output	2.8	–4.4	5.2
Advanced Economies	1.7	–5.8	3.9
Emerging Market and Developing Economies	3.7	–3.3	6.0

Source: IMF, 2020

group of emerging market and developing economies is projected to contract by –3.3 percent in 2020 and expected to grow by 6.0 percent in 2021. There are the uneven recoveries over the period of time during the year. It shows that the advanced economies which mostly practice neoliberalism have immense contraction and flat recovery- a sign of poor resiliency economy. However, it is notable that China with the state capitalism could keep its economy in positive side growing even in this great recession.[17] Nevertheless, the growth projections of IMF for 2021 seems unachievable since the revival of the economy in such a speed is impossible until and unless there are very focused fast-track interventions.

The global impact of the COVID-19 pandemic is still expanding daily. The cumulative output loss to the global economy across 2020 and 2021 from the pandemic crisis will be over $12 trillion (IMF 2020). Moreover, there are some countries suffering from multiple epidemics, i.e., the Coronavirus, Locust invasion and the protracted conflicts at a time. Governments around the globe are confronted with multiple challenges related to minimizing the devastating health impact and protecting human lives, and ensuring sufficient food supplies and the functioning of services to those most in need (WESP 2020). The WHO has already announced to take precautionary measures like social distancing until 2022.

2.3 Emergence of New Policy

In the course of human civilization, Adam Smith conceptualized the "laissez-faire" economy back in 18th century that advocates "free" market and anti-regulatory policies, politically termed as the capitalism. Even though there was counter policy of the "planned economy" under the communism-oriented regime the "free market" kept its dominant role globally till the date. With the Great Depression (1930s) and, then, World War II (1939–1945),

[17] https://blogs.imf.org/2020/10/13/a-long-uneven-and-uncertain-ascent/

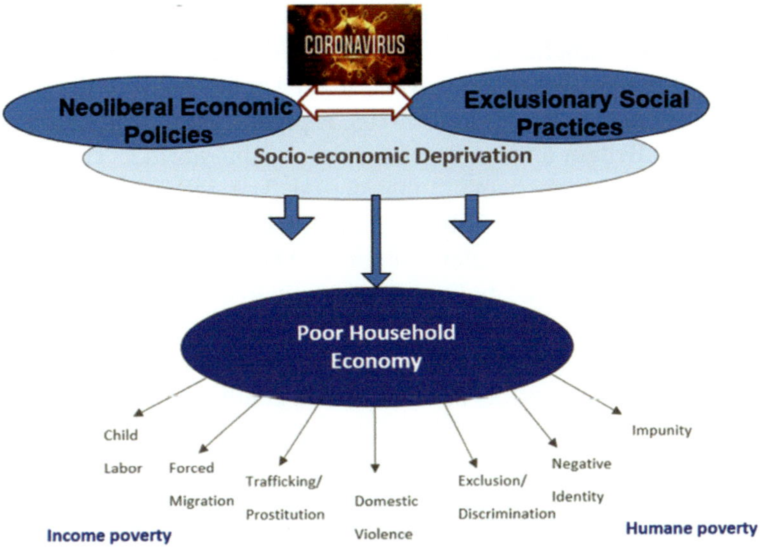

Figure2.6 Systemic marginalization.
Source: Author

it was believed that governments needed to play a role in regulating their economies and that the laissez faire paradigm could not operate with only an "invisible hand." A real hand needed to serve as a countervailing element to control its cycles and any speculative abuses.[18] There were some economists like John Maynard Keynes who modified the policy to make nation-state, the real hand, more responsible as a market driver so as to build welfare state. The Keynesian economics evolved during the great depression of 1930s which advocated for increased government expenditures and lower taxes to stimulate demand and pull the global economy out of the depression.[19] In fact, Keynes proposed that governments must act as compensatory agents at all times to ensure full employment by way of a broad program of discretionary fiscal policy, which checks and balances every aspect of the capitalist economy.[20] With this the capitalist crisis with its shrinking profit rates inspired

[18] https://www.jussemper.org/Resources/Economic Data/The Neo-Capitalist Assault/Resources/KeynesianEconomics.pdf

[19] https://www.investopedia.com/terms/k/keynesianeconomics.asp

[20] https://www.jussemper.org/Resources/Economic Data/The Neo-Capitalist Assault/Resources/KeynesianEconomics.pdf

the corporate elite to revive economic liberalism that made it "neo" or new,[21] then it became neo-liberal economic policy and spread widely as the globalization. The then British premier Margaret Thatcher became the pioneer of neoliberalism policies in the mid-1970s, a period when Keynesian policies still reigned in Britain and much of the rest of the world.[22] The policy was very famous by "Thatcherism" at that period. The policy has been imposed through powerful financial institutions like the International Monetary Fund, the World Bank, and the Inter-American Development Bank. However, the Marxists disparage the policy as the capitalists adopted the neoliberal approach because their class power had been diluted under Keynesianism and was threatened in the mid-1970s.[23] They took neoliberalism as a measure to restore the class power because it creates condition for class formation. During the cold war period (1947–1991), many countries adopted "mixed economy" to pretend themselves non-aligned to the super powers: NATO and Warsaw Pack. Even though the principle of mixed economy was to bring the best practices of the two economic models and make the economy prosperous, no "third word" countries, except very few, could do so rather many countries fell into the protracted conflicts that again degraded their economy. In the name of globalization free market policy became the means of contemporary form of imperialism through invasion in the national economy damaging domestic/local systems. Even in its early stage Marx predicted that the capitalist economies would have periodic crises and ultimately the system would collapse (Ahuja 1992). In fact, the great depression of 1930 was the result of free market, the financial crisis of 2008/09, the result of neoliberalism and the recent COVID-19 pandemic, to some extent, the result of globalization.

Joseph Stiglitz (2019), Nobel laureate in economics for 2001, argued that our economic and political systems are rigged against the middle class and in favor of the wealthy. He explains how the American economy—and its government—tilts toward big business, a few corporations have come to dominate entire sectors of the economy, contributing to skyrocketing inequality and slow growth. Too many have made their wealth through exploitation of others rather than through wealth creation. Thus, the neoliberalism made the space for exploitation of natural resources (public wealth) and human resources for

---

[21] http://web.inter.nl.net/users/Paul.Treanor/neoliberalism.htmlhttp://web.inter.nl.net/users/Paul.Treanor/neoliberalism.html

[22] https://www.theguardian.com/global-development/poverty-matters/2013/apr/16/legacy-margaret-thatcher-neoliberalism

[23] https://www.marxists.org/archive/harman/2008/xx/neolib.htm

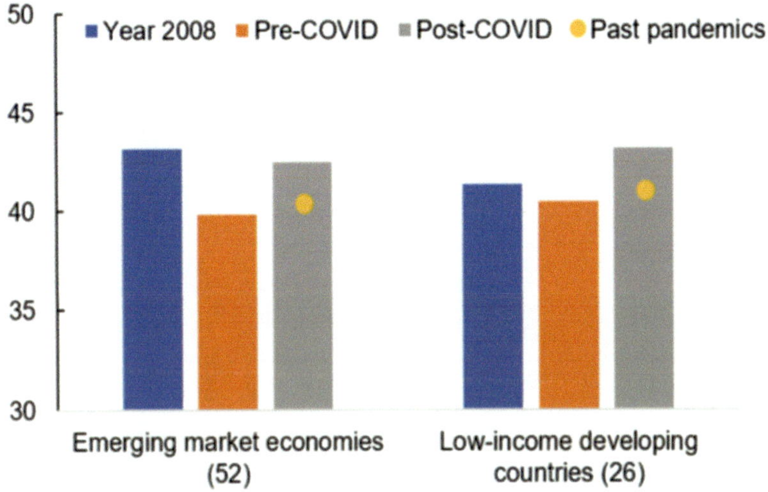

Figure 2.7 Reversal of gains on inequality.
Source: IMF, 2020

the sake of profit (private gain) to the limited persons-profiteering character of the free market mechanism. They have no concern to disparity, poverty, carbon-emission, and or environment degradation. The division of labor as provoked by Alfred Marshall has been turned into the division of laborer that generated the caste, class, or xenophobia, the "neoliberal atrocities" (Bk 2018). The division of people on the basis of their birth which sustained inter-generational. The excluded ones always remain in the persistent vicious circle of deprivation. The Coronomics,[24] economic consequences due to COVID-19 pandemic, has further intensified that slur,[25] widened the inequality of opportunities and unemployment shock pushing many middle-class people into the poverty. The estimated effect from COVID-19 on the income distribution is much larger than that of past pandemics. The analysis shows that the average Gini coefficient for emerging market and developing economies will rise to 42.7, which is comparable to the level in 2008 with the impact larger for low-income developing countries.[26] This widening inequality on average has

[24] https://www.academia.edu/42328909/Revival_of_Global_Economy_after_COVID-19_Pandemic

[25] https://edition.cnn.com/2020/04/15/asia/india-coronavirus-lower-castes-hnk-intl/index.html

[26] https://blogs.imf.org/2020/10/29/how-covid-19-will-increase-inequality-in-emerging-markets-and-developing-economies/?utm_medium=email&utm_source=govdelivery

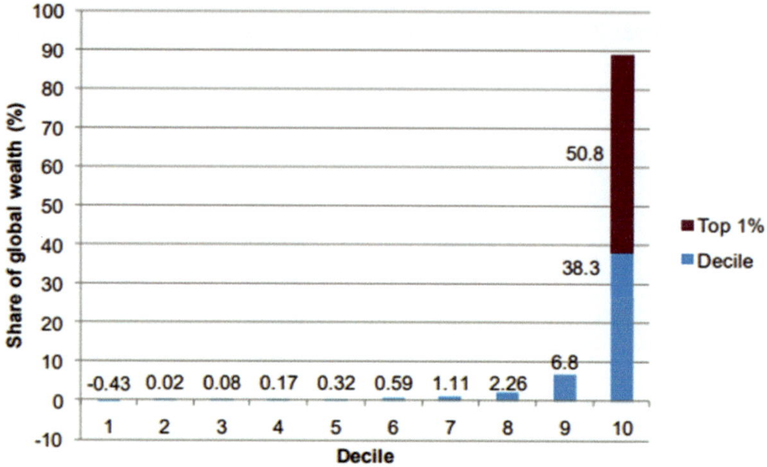

Figure 2.8 Global distribution of wealth by decile.
Source: OXFAM, 2017

a clear impact on people's well-being. It could reduce welfare by 8 percent in emerging markets and developing countries with more than half of it stemming from the excess change in inequality as a result of a person's ability to work from home. Nobel laureate for 1998, Amartya Sen understands "poverty" as deprivation in the capability to live a good life, and "development" as capability expansion in his capability approach.[27] So, poverty is not caused by a person's unwillingness to work hard or lack of skill. As a matter of fact, a poor person works very hard and has more skill. S/he languishes in poverty because s/he does not receive the full worth of the work.[28] Marx, in his labor theory of value (Ahuja 1992), describes it as the rate of exploitation (the ratio of profit share to the wage share). According to him, the forces at work in a capitalist economic system bring about the increase in the rate of exploitation of labor. Under the existing social and economic institutional arrangements, someone else always comes in-between and takes away a significant part of income leaving less to those who worked hard (CGAP 2003). The existing economic institutions are designed in such a way that it allows this process to continue and gather strength everyday so that the earnings of others can

---

[27] https://www.iep.utm.edu/sen-cap/

[28] https://www.academia.edu/33263834/Social_Inclusion_through_Microfinance_A_Bottom_Up_Approach

make a handful of people richer every day and turn large number of people into paupers (Giddens 1998). The OXFAM report (2017) shows the richest one percent of the world population is worth more than the other 99 percent based on the false assumptions of neoliberalism. OXFAM calls it "Crony capitalism" in its report which benefits the rich at the expenses of the common good and poverty reduction. So, people dividing into the economic class is the outcome of the economic liberalization. The UN revealed a sense that inequalities were growing and that small elites were benefiting from development and growth at the expenses of the majority (UN 2013). So, poor are not poor themselves rather they have been made poor through the systemic structural barriers created previously with feudalism and latter perpetuated with neo-liberalism.[29] Under this economic policy "winner-takes-all" markets are structural and it simply ignored the social problems produced by deregulated markets, which created serious threats to social cohesion (Giddens 1998). Abhijit Banerjee and Esther Duflo (2011), the Nobel Prize winning couple for 2019, describe that the universal traps of ignorance, ideology, and inertia often stymie policies and institutions. They believe that small changes can have big effects. However, since these barriers have been crafted skillfully over the centuries to benefit a handful of people it is hard to bring emancipatory shift without having substantial intervention.

In conclusion, capitalism has generated massive wealth for some, but it has devastated the planet and has failed to improve human well-being at scale.[30] Corporate capitalism is committed to the relentless pursuit of growth, even if it ravages the planet and threatens human health. So, we need to build a new system that will balance economic mobility and human flourishing. The concept of "Protective Liberalism" (Bk 2018) has been introduced as a replacement to libertarianism that is the socio-economic policy for "Liberal Socialism."

2.4 How Protective Liberalism or Liberal Socialism Resets Coronomics

Neoliberalism could not efficiently balance the global economy, rather it created the environment of the rich grow richer and the poor grow poorer around the world there has been already some structural modification on the policy

[29] https://www.academia.edu/33263834/Social_Inclusion_through_Microfinance_A_Bottom_Up_Approach

[30] https://www.forbes.com/sites/drewhansen/2016/02/09/unless-it-changes-capitalism-will-starve-humanity-by-2050/

with protectionism after the financial crisis of 2008–09.[31] The social inequalities among the people are the consequence of neo-liberal economy or capitalism which is against egalitarianism; so the equality of opportunity is seen as desirable and necessary. For that a welfare state needs a strong system of solidarity and protection to its weaker section of the beneficiaries such as the vulnerable, the disadvantaged, and the left-outs.[32] The state policy of equitable economic mobility and social entrepreneurship through protective liberalism would support and promote not only achieving the goal of liberal socialism but also restore the egalitarian society in a sustainable manner. It seems that free market is the ideal force of stimulus at which there is no or less state for market mechanism whereas the protective liberalism is the practical solution within the frame of more responsible "welfare state" under the political thought of liberal socialism. Mickey Kaus has evolved a concept of "civic liberalism" (Giddens 1998) similar to protective liberalism. Likewise, Oskar Ryszard Lange (1936–37) evolved a similar concept of Market socialism—a type of economic system involving the public, cooperative, or social ownership of the means of production in the framework of a market economy (Sen 1997). It is a theoretical concept (model) of an economic system in which the means of production (capital) are publicly or collectively owned, and the allocation of resources follows the rules of the market (product-, labor-, capital-markets). In this economic system, he includes "distribution according to relative needs" which is very close to protective measure to the deprived ones. Likewise, Anthony Giddens (1998) blended two opposing ideologies rationally and brought "the Third Way" of renewing "Social Democracy." The "third way" suggests that it is possible to combine social solidarity with a dynamic economy. To pursue it, we need "less national government, less central government, but greater governance over local processes," as well as opening out in the direction of the global community.

Since the Corona-outbreak originated in China, a country with state capitalism, its economic openness is somewhat responsible for its spread.[33] State capitalism is a system where the state dominates the market's activities for a political purpose. Adolf Wagner (1837–1917) regarded state capitalism as a solution of compromise between competitive capitalism and socialism (Elena 2012). It is a system in which the governments realized that profitable

---

[31] http://web.inter.nl.net/users/Paul.Treanor/neoliberalism.html

[32] https://www.academia.edu/41584548/Protective_Liberalism_The_Foundation_for_Transformative_Democracy

[33] https://www.academia.edu/42328909/Revival_of_Global_Economy_after_COVID-19_Pandemic

enterprises owned by the state make the state stronger. Adolf Wagner has foreseen a system of redistribution of wealth between the classes of society with the help of the fiscal system, meant to stop people from gaining too much wealth. However, state capitalism in China and in the rest of the world has several limitations, i.e., the state-owned companies are less efficient, use cheap capital, throw aside the private entrepreneurs, comparatively have less safety-nets and benefit for political favoritism.[34]

The COVID-19 pandemic has now strongly justified that the nation-state should be a responsible welfare state. The welfare system should be restructured in an innovative way. Social innovation frequently occurs at the local level; communities as a part of larger civil society have their indigenous skill, knowledge, and practices that can be base to dig in for innovation. Social welfare is the result from a change in economic and non-economic variables and a function of the utility (satisfaction) level of all individuals constituting the society. But any changes may benefit one section of the society and harm the other. For that, Kaldor–Hicks–Scitovsky developed the compensation principle.[35] However, when the harms go on for centuries in some caste, class, or race of people, then the state should satisfy them with reparation policy for their lost opportunities and dignity. Once the social choice (Ahuja 1992), i.e., neoliberalism is made, all the individual comprising the society are compelled to accept it, since social preference dominates individual preferences. It is not because they would benefit from it, but because they do not have an alternative social choice for the time being, no individual can disregard the social choice derived through a process. They would rather have deprived of their capability resulting the inequality of opportunities. Amartya Sen proposed an inequality-adjusted welfare function as the average per capita income of a measured group, y (e.g., nation) is multiplied with (1-G) where G is the Gini index; $W=y(1-G)$, a relative inequality measure.[36] Nevertheless, the inequality is the result of exclusion, a social injustice. So, the welfare should be measured with the inclusion/exclusion dynamics of the policy. The welfare depends on efficiency of the policy and it should be oriented toward the "Pareto Optimality." The research (Bk 2008) revealed that higher the inclusivity of a policy, higher is its utility (welfare).

The reparation policy as a protective measure should be undertaken to compensate the historical loss in welfare. It should be a righteous phenomenon

[34] https://www.academia.edu/41584548/Protective_Liberalism_The_Foundation_for_Transformative_Democracy

[35] *H.L. Ahuja (1992), Advanced Economic Theory, S. Chanda and Company Ltd, New Delhi.*

[36] https://en.wikipedia.org/wiki/Social_welfare_function

in a democratic "welfare state." So, the sustained protective measures in the market mechanism not only compensate the loss or harm due to the social choice but also redistribute the income in an equitable manner. It would bring the emancipatory shift (Bk 2018) to those marginalized section of the society. Therefore, while making a welfare policy, the state needs to analyze its losers and simultaneously establish a compensation policy as the protective measure which is the main ingredient of protective liberalism. However, the question always raises on the process of making social choice or judging social welfare. Simple way of dealing it is the inclusionary representation policy as characterized with transformative democracy, the pillar of liberal socialism. It is beyond private or corporate capitalism and state capitalism at which the society or even a community will have command over the means of production, ultimately for the social wellbeing. So, it can be coined as the "social capitalism" as well.

Marx opined that the capitalists are able to exploit the labor class and extract surplus value from it because they have the ownership of non-human means of production (Ahuja 1992). Amartya Sen in his entitlement approach marked that an entitlement set of a person is determined by his original bundle of ownership and additional bundles acquired; and famine is seen through the lens of loss of such entitlement.[37] These days, the equity of the most of the corporate houses is concentrated in the hands of a selected few. Those large number of absent owners have no stake in the communities in which the companies operate.[38] So, the collective ownership or social entitlement through the stake-holding that widens the proximity of redistribution of surplus value is the basic of the policy. Likewise, the compensatory provisions within progressive taxation are other fiscal mechanisms to make it more equitable. Building inclusive governance from the bottom is the hardcore strategy of the liberal socialism. It assures both redistribution of income and increased social solidarity. It involves a balance between regulation and deregulation at national and sub-national levels; and a balance between the economic and the non-economic variables in the life of the society.

(Dr. Bk is a former Fulbright Research Scholar at Brandeis University in the USA, the author of the book "Eradicating Hunger: Rebuilding Food Regime," visiting Faculty member for MBA-BF Program of Tribhuvan University and the Advisory Editorial Board Member for the International

---

[37] https://nkfood.wordpress.com/2012/08/31/summary-of-food-economics-and-entitlements-by-amartya-sen/

[38] https://www.forbes.com/sites/drewhansen/2016/02/09/unless-it-changes-capitalism-will-starve-humanity-by-2050/

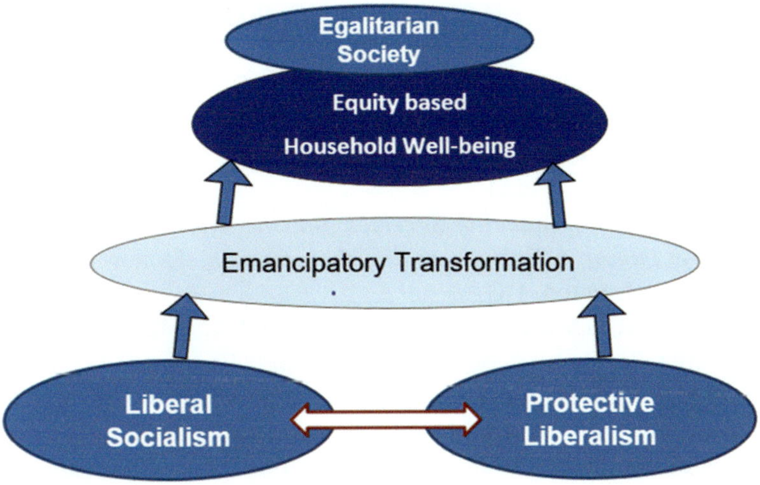

Figure 2.9 Systemic transformation.
Source: Author

Journal of Commerce and Management Studies, India. He is now working as the Principal Secretary for Provincial Government of Bagamati, Nepal.)

Note: Paper presented in the Fifteenth Annual Himalayan Policy Research Conference 2020 organized by the University of New Mexico, the USA on 5th December, 2020.

References

1. Ahuja, H.L. (1992), *Advanced Economic Theory,* S. Chanda and Company Ltd, New Delhi.
2. Banerjee, Abhijit and Duflo, Esther (2011),*'Poor Economics: A Radical Rethinking of the Way to Fight Global Poverty,* Public Affairs, NY
3. Bhandari, Medani. P. (2020). In the Covid-19 Regime – What Role Intellectual Society Can Play. International Journal of Science Annals, 3(2), 5–7. doi:10.26697/ijsa.2020.2.1 https://ijsa.culturehealth.org/en/arhiv https://ekrpoch.culturehealth.org/handle/lib/71
4. Bhandari, Medani P. (2020), Editorial- the Role of Scholarly World – Science Writing, Scientific Journal of Bielsko-Biala School of Finance and Law, ASEJ 2020; 24 (2): 4-4, https://asej.eu/resources/html/article/details?id=207187
5. Bhandari, Medani P. (2020), The Phobia Corona (COVID 19) - What Can We Do, Scientific Journal of Bielsko-Biala School of Finance and

Law, ASEJ 2020, 24 (1): 1-3, GICID: 01.3001.0014.0769, https://asej.eu/resources/html/article/details?id=202946
6. Bk, Man Bahadur (2018), *Eradicating Hunger: Rebuilding Food Regime,* EKTA Books, Nepal
7. Bk, Man Bahadur (2008), *Social Inclusion in Microfinance,* Jana UtthanPratisthan, Nepal
8. CGAP (2003), *Financial Services for the Rural Poor: Helping to Improve Donor Effectiveness in Microfinance*, Donor Brief No. 15, CGAP, Washington, DC.
9. Elena, PAGUBA Cristina (2012), "Liberal Capitalism Verses State Capitalism", Faculty of Economics and Business Administration/ Department of Political Economics, Babes-Bolyai University, Cluj-Napoca, Romania. Available at https://www.ubbcluj.ro/
10. Giddens, Anthony (1998), *The Third Way: The Renewal of Social Democracy,* Blackwell Publishers, USA.
11. ILO (2020), *ILO Monitor: COVID-19 and the world of work. Third edition Updated estimates and analysis, ILO, 2020.*
12. IMF (2020), *World Economic Outlook October 2020: A Long and Difficult Ascent,* International Monetary Fund.
13. Lal, B Suresh (2020), *Health and economic shocks of COVID-19,* International Journal of Multidisciplinary Research and Development, Volume 7; Issue 4; 2020 available at www.allsubjectjournal.com.
14. OXFAM (2017) available at https://www.oxfamamerica.org/explore/research-publications/an-economy-for-the-99-percent/
15. Sen, Amartya (1981), *Poverty and Famine: An Essay on Entitlement and Deprivation,* Clarendon Press-Oxford, New York.
16. Sen, Amartya (1997), *On Economic Inequality,* Oxford University Press, New York.
17. Stiglitz,Joseph(2019),People,Power,andProfits:ProgressiveCapitalismfor anAgeofDiscontent,availableathttps://www.marketplace.org/2019/05/30/economist-joseph-stiglitz-progressive-capitalism-people-power-profits/
18. UN (2020), *Shared Responsibility, Global Solidarity: Responding to Socio-economic Impact of COVID-19*, United Nations
19. UN (2013), *Inequality Matters: Report on the World Social Situation-2013,* Department of Economic and Social Affairs, Population Division, United Nations.
20. WB (2020), *Fragility and Conflict: On the Frontline against Poverty,* The World Bank Group.
21. WESP (2020), *World Economic Situation and Prospects 2020,* United Nations.

3

On the Unified Index for Balanced Development and Inequalities Level Estimation

Sergii Polumiienko

Doctor in Physics and Mathematics, Institute of Telecommunications and Global Information Space of the National Academy of Sciences of Ukraine, 13, Chokolivsky Blvd, Kyiv, 03186, Ukraine serge_pol@ukr.net

ABSTRACT

The paper deals with an approach aimed at constructing balanced development strategies for the country and its regions, including overcoming inequalities. There are defined regional (within the country) resources and development is considered as a set of system participants strategies aimed at creation of new resources using available information. The categories, types, and sets of basic indicators of regional resources are introduced.

A cooperative resource model of balanced development is being constructed. This development is considered as the result of the strategies determined by the Shapley value. But such strategies may not meet normal development; they depend on state region, which must be balanced at first, because balanced strategies will reproduce initial state. The strategies coalition enters a game with a certain number of resources and expects the corresponding them sharing. In practice, such sharing usually is not performed. The model also indicates that in order to bring the region to a balanced use of resources, unbalanced within cooperative model strategies should be used. As a result, it is proposed to define the regional development index.

Comparative estimations of regional resources are introduced. In addition to quantitative, qualitative estimations of resource types are also considered It is proposed a unified development index. This index and indicators allow defining the balance of current and target state of regions and country, as well as strategies for their achievement. The same tools allow evaluating and comparing factors of inequality, and on this basis finding constructive strategies for overcoming those factors.

Keywords: inequality, unified development index, balanced development, indicative analysis, game-theoretical model.

3.1 Introduction

Achieving development efficiency, obtaining maximum income while minimizing the cost of various resources today, as never before, require the use of a variety of specially created technologies, first of all, information technologies. The specialized tools considered as they can be used to achieve the goals, to create new resources at the minimal expense..

The definition of development at the end of the last century was given a new concept—"sustainable development," which was defined as "development that meets the needs of the present, without compromising the ability of future generations to meet their own needs" [1]. In Germany, however, the sustainable development roots reach the middle of the last century [2]. But even though the number of publications on sustainable development has exceeded 150,000 [3], there are no significant changes in this direction. The fact is that sustainable development in the accepted definition contradicts the current tendencies of economic activity, the existing principles of its effectiveness, the concept of unlimited growth as noted in [4] and many other publications. All this has obviously led to the UN transition from the initial definition of sustainable development to the approval of its 17 goals. The tenth of these goals is "Reduced Inequalities" [5].

At the same time, many indicator systems and integral indexes, which have their tasks and methods of calculation, have been created to determine the sustainable development level. The analysis of these materials [6], [7], however, did not allow us to determine a universal approach to assess the state of available resources, capable of providing objective estimates for different countries and their regions.

The purpose of this paper is to present a unified approach to resource estimation as a basis for formulating strategies for balanced development of regional (within the country) and national systems and, first, to reduce inequalities within them. The need for new approaches for estimating and responding to the inequalities of human development is also emphasized in [8].

3.2 Cooperative Resource Model of Balanced Development

Let us first consider construction of a unified index of regional and national systems, including, balanced development. We will use this index methods to construct further the inequality index and methods of reducing it.

Basing on the importance of the region's specifics, their resources, which differ significantly in Ukraine and many other countries of the world, what was noted, for example, for the Seoul regions [9], we will proceed from the country regions n, $n = 1,..., N$, denote the volume of their resources available at the moment of time $t \in [t_0,T]$ by $res_{n,l,k,m}(t)$, where l corresponds to the category, $l = 1,..., L$, k to the type, $k = 1,.., K(l)$, and m—to the specific resource, $m = 1,..., M(k)$,

There are considered categories of natural, economic, social, infrastructural, informational, and research and development resources (see Table 3.2 and [10]). The categories include about 45 types of resources, which are described by more than 450 basic indicators.

Let us define comparative shares of resources

$$r_{n,l,k,m}(t) = res_{n,l,k,m}(t) / \sum_n res_{n,l,k,m}(t)$$

in their national volume. These values can be considered also as the part of the global volume of n,l,k,m-resource. These comparative shares probably form the most objective quantitative estimates of the available resources.

We will identify the participants of the regional socio-ecological-economic system as players $i_{n,l,k,m}$, controlling the resource $r(i_{n,l,k,m},t)$. We will assume that players' actions are aimed at increasing the resources they match. Let us call these actions strategies and denote by $s(i_{n,l,k,m},t)$, $s(i_{n,l,k,m},t) \in S(i_{n,l,k,m},t)$, $S(i_{n,l,k,m},t)$ — the set of strategies of the player $i_{n,l,k,m}$ at moment of time t.

To increase the resource $r(i_{n,l,k,m},t)$, the player uses a certain set of other resources. At the same time, the player provides $r(i_{n,l,k,m},t)$ to other players to execute their strategies. Also, $s(i_{n,l,k,m},t)$ is influenced by other players' strategies aimed at $r(i_{n,l,k,m},t)$. To reduce this impact, as well as to increase his capabilities, the player must cooperate with other players.

We will call the coalition $c_{n,l,k,m}$ the union of players i, that have an agreed (identical) strategy $s(c_{n,l,k,m},t)$, $c \in C$, C—the set of all coalitions [11]. We will only consider coalitions and assume that players $i_{n,l,k,m}$, fulfilling the strategies $s(i_{n,l,k,m},t)$, enter the coalition $c_{n,l,k,m}$, which controls the resource $r_{n,l,k,m}$ with the part $rc_{n,l,k,m}(s(i_{n,l,k,m},t))$ of its n,l,k,m-resource.

The amount of resources provided by players creates a joint resource $r_{n,l,k,m}(s(c_{n,l,k,m},t))$ of the coalition $c_{n,l,k,m}$, with which it executes its strategy $s(c_{n,l,k,m},t)$

$$rc_{n,l,k,m}(s(c_{n,l,k,m},t)) = \sum_{i_{n,l,k,m}\in c_{n,l,k,m}} rc_{n,l,k,m}(s(i_{n,l,k,m},t)). \qquad (3.1)$$

When fulfilling a strategy, players $i_{n,l,k,m} \in c_{n,l,k,m}$ spend (parts of) resources $rd_{n,l,k,p}(s(i_{n,l,k,p},t))$, $p\neq m$, which are controlled by other players $N\{i_{n,l,k,m}\}$, $m,p = 1,..,M$, which in the sum are

$$rd_{n,l,k,m}(s(c_{n,l,k,m},t)) = \sum_{\substack{i_{n,l,k,p}\in c_{n,l,k,p}\\ p=1,p\neq m}}^{M(k)} rd_{n,l,k,p}(s(i_{n,l,k,p},t)) \;. \qquad (3.2)$$

Thus, at moment of time t_{j+1} a coalition $c_{n,l,k,m}$ in order to increase its resource $r(c_{n,l,k,m},t_j)$ adds to it $rc_{n,l,k,m}(s(c_{n,l,k,m},t_{j+1}))$ and spends $rd_{n,l,k\,m}(s(c_{n,l,k,m},t_{j+1}))$ -

$$\begin{aligned} &r_{n,l,k,m}(s(c_{n,l,k,m},t_{j+1})) = r_{n,l,k,m}(s(c_{n,l,k,m},t_j)) + \\ &+rc_{n,l,k,m}(s(c_{n,l,k,m},t_{j+1})) - rd_{n,l,k,m}(s(c_{n,l,k,m},t_{j+1})), \end{aligned} \qquad (3.3)$$

where $r_{n,l,k,m}(s(c_{n,l,k,m},t_j))$ - is a value of n,l,k,m- resource at moment t_j.

The resource $r_{n,l,k,m}(s(c_{n,l,k,m},t_{j+1}))$ corresponds to one type of coalition activity. But other coalitions $c_{n,l,k,p} \in C$, $p\neq m$, also change the n,l,k,m-resource.

Let's call the situation [11] the vector of strategies of all coalitions $c_n \in C_n$ and denote it by $sit_n(t)$,

$$sit_n(t) = (s(c_{n,1,1},t)),.., s(c_{n,L,K,M},t)),$$

and through $Sit_n(t)$—the set of all situations $sit_n(t)$ occurring in region n.

In the situation $sit_n(t_{j+1})$,), by analogy with (3.1) – (3.3), the resource value $r_{n,l,k,m}(t_{j+1})$ is defined as

$$r_{n,l,k,m}(sit(t_{j+1})) = r_{n,l,k,m}(sit(t_j)) + rc_{n,l,k,m}(sit(t_{j+1})) - rd_{n,l,k,m}(sit(t_{j+1})),$$

where $rc_{n,l,k,m}(sit(t_{j+1}))$ and $rd_{n,l,k,m}(sit(t_{j+1}))$ expresses accordingly the part of the resource created by the coalition $c_{n,l,k,m}$ and spent by all other coalitions $c_{n,l,k,p} \in C_n$, $p\neq m$, fulfilling their strategies.

The coalition interest $c_{n,l,k,m}$ to increase the n,l,k,m-resource volume, thus, is to increase the difference

$$rc_{n,l,k,m}(sit(t_{j+1})) - rd_{n,l,k,m}(sit(t_{j+1})), \tag{3.4}$$

influenced by strategies of other coalitions from C_n. This is in line with current development trends, which can be defined as maximization of (3.4) by separate coalitions. The harmonization of the interests of the entire society, i.e., its sustainable development, requires the coordination of strategies not of separate but of all coalitions $c_n \in C$. This is clearly contrary to maximization of (3.4).

After the situation is performed, the coalition claims to receive payoff, which depends on the respective changes of all resources, and is a share created by all coalitions. This payoff depends on the contribution made by the coalition $c_{n,l,k,m}$ strategy, and on the coordination of all coalitions strategies on the n,l,k,m-resource, which results from the interaction of all coalitions across all resources.

The function v, which matches to the coalition C, the greatest confidently received payoff $v(C)$ (division $x = (x_1,..,x_n)$), is called the characteristic function [11], [12], and the set $<I, v(C)>$—a characteristic function form cooperative game,

$$x_i \geq \upsilon(i) \text{ for } i \in I \text{ and } \sum_{i \in I} x_i = v(I),$$

The coalition $c_{n,l,k,m}$ payoff is defined by the following function:

$$v(c_{n,l,k,m}, t_j) = \max_{S_{c_{n,l,k,m}}(t_j)} \min_{S_{C_n \setminus c_{n,l,k,m}}(t_j)} r_{n,l,k,m}(sit(t_j)) = \max \min_{Sit_n(t_j)} r_{n,l,k,m}(sit(t_j)). \tag{3.5}$$

Thus, a coalition payoff is the maximum resource value that $c_{n,l,k,m}$ can receive, basing on the set $Sit_n(t_j)$ of situations and own strategies $S_{c_{n,l,k}}(t_j)$.

Function (3.5) can be defined as a characteristic function of the game:

$$\Gamma_n(t_j) = <C_n, Sit(t_j), v(c_{n,l,k}, t_j), n = 1,.., N, l = 1,,., L, k = 1,,., K, t_j \in [t_0, T]>, \tag{3.6}$$

which defines the coalitions interaction to create and use resources in region n.

Game (3.6), including (3.1) – (3.5), is called the cooperative resource model of the region.

Optimal strategies in the model were defined by Shapley value, which indicates the optimal division in (3.6) and characterizes what a coalition $c_{n,l,k,m}$ can achieve upon completion of the game when $c_{n,l,k,m}$ interests and strategies are coordinated with other coalitions. This division also defines the balance of resources in the situation $sit(t_j)$. Appropriate strategies are also defined as balanced development strategies.

The division of the cooperative model provides for the payment to coalitions of the share created by all participants in accordance with the resources invested by them in the implementation of the coalition strategy. But this division has little regard for the irreversible consumption of resources, especially in the case of full resource expenditure, which occurs, for example, in the extraction of minerals.

In addition, these situation optimality conditions, and their corresponding behavior may in practice not correspond to normal, including balanced development, since the constructed model does not consider the real strategies content. Shapley's value-based state of resources and optimal strategies can even reflect the balanced destruction of a system due to the majority of destroying components in their strategies. In the case of unsatisfactory condition of social and economic resources, the balanced development strategies, constructed using the model, will preserve this state, that is, prolong the stagnation of the region.

Thus, the constructed model presented above allow us to recognize formal basics as for the absence of balanced development strategies, so as for the complexity of cooperative interaction in development of all resources, also the need of other ways of balanced development defining, including overcoming inequality.

3.3 Unified Index of Regional and National Development

Let's introduce the values

$$r_{l,k,m}^{\min}(t_j) = \min_n r_{n,l,k,m}(t_j),\ r_{l,k,m}^{\max}(t_j) = \max_n r_{n,l,k,m}(t_j),$$

—minimum and maximum in all regions of the country n value of the comparative estimates of the resource $r_{n,l,k,m}(t_j)$ of type k and category l at the moment of time $t_j \in [t_0, T]$.

The value

$$rm_{l,k,m}(t_j) = r_{l,k,m}^{\max}(t_j) - r_{l,k,m}^{\min}(t_j),$$

then indicates the maximum variation of values $r_{n,l,k,m}(t_j)$, and value

$$rr_{n,l,k,m}(t_j) = r_{n,l,k,m}(t_j) - r_{l,k,m}^{\min}(t_j), \text{ —}$$

the relation of the resource volume $r_{n,l,k,m}(t_j)$ and its minimum value at time t_j by regions n.

The relation

$$pr_{n,l,k,m}(t_j) = \frac{rr_{n,l,k,m}(t_j)}{rm_{l,k,m}(t_j)} = \frac{r_{n,l,k,m}(t_j) - r^{\min}_{l,k,m}(t_j)}{r^{\max}_{l,k,m}(t_j) - r^{\min}_{l,k,m}(t_j)}, \tag{3.7}$$

indicates the comparative value of the resource volume of type k of category l, normalized by the variation of values $r_{n,l,k,m}(t_j)$ at time t_j by regions n.

The approach to such definition of different estimates is used in some indices of the development level of socio-ecological-economic systems.

Equation (3.7) does not take into account the positivity and negativity of resources. We will then define by (3.7) the positive resources and the negative, for example, solid waste, by the value

$$dr_{n,l,k,m}(t_j) = 1 - \frac{r_{n,l,k,m}(t_j) - r^{\min}_{l,k,m}(t_j)}{r^{\max}_{l,k,m}(t_j) - r^{\min}_{l,k,m}(t_j)} = \frac{r^{\max}_{l,k,m}(t_j) - r_{n,l,k,m}(t_j)}{r^{\max}_{l,k,m}(t_j) - r^{\min}_{l,k,m}(t_j)}. \tag{3.8}$$

Then the following difference expresses the estimate of the resource state

$$rc_{n,l,k,m}(t_j) = pr_{n,l,k,m}(t_j) - dr_{n,l,k,m}(t_j) \tag{3.9}$$

where $pr_{n,l,k,m}(t_j)$ and $dr_{n,l,k,m}(t_j)$ reflects positivity ($dr_{n,l,k,m}(t_j) = 0$) and negativity ($pr_{n,l,k,m}(t_j) = 0$) of n,l,k,m-resource at the moment of time t_j, value

$$rc_{n,l,k}(t_j) = \sum_{m=1}^{m(K)} (pr_{n,l,k}(t_j) - dr_{n,l,k}(t_j)), \tag{3.10}$$

—comparative estimates of the resource type k state, and

$$rpd_{n,l,k}(t_{j+1}) = rc_{n,l,k}(t_{j+1}) - rc_{n,l,k}(t_j),$$

—change of n,l,k,m-resource at the moment of time t_{j+1} due to the activity of all members of the system.

In addition to quantitative estimates, we will consider qualitative estimates that reflect the properties of resources, including characteristics of access and pressure on them, the potential for their use and recovery, and more. In [10] there are considered the qualitative estimates corresponding to each of the resource types. Because of their large number below the set of generalized estimates for use simultaneously for all types of resources is proposed (Table 3.1).

Table 3.1 Generalized qualitative estimates

	Properties of resource types	Current usage trends and resource status	Expected status according to the current trends
1	Availability and quality	$q_{n,l,k,1}$	$qv_{n,l,k,1}$
2	Pressure, need for internal and external members, regions, countries	$q_{n,l,k,2}$	$qv_{n,l,k,2}$
3	Impact on other resources, regions and countries	$q_{n,l,k,3}$	$qv_{n,l,k,3}$
4	Use intensity	$q_{n,l,k,4}$	$qv_{n,l,k,4}$
5	New technologies development and application	$q_{n,l,k,5}$	$qv_{n,l,k,5}$
6	Safety, vulnerability to damage, harmful use	$q_{n,l,k,6}$	$qv_{n,l,k,6}$
7	Recovery and development, their potential, including as an independent	$q_{n,l,k,7}$	$qv_{n,l,k,7}$
8	Perception of the current state by society	$q_{n,l,k,8}$	$qv_{n,l,k,8}$
9	Attitudes in society towards its other members, collective resources	$q_{n,l,k,9}$	$qv_{n,l,k,9}$
10	Society resistance to the conflicting interests of members, regions, countries	$q_{n,l,k,10}$	$qv_{n,l,k,10}$
11	World-wide relevance, role in the international market	$q_{n,l,k,11}$	$qv_{n,l,k,11}$
12	Effectiveness of measures to improve current trends	$q_{n,l,k,12}$	$qv_{n,l,k,12}$
	Total estimation	$q_{n,l,k} = \sum_{i=1}^{12} q_{n,l,k,i}$, $qv_{n,l,k} = \sum_{i=1}^{12} qv_{n,l,k}$ $q_{n,l,k,i} = 1,2,..,10.$ $qav_{n,l,k} = q_{n,l,k} + qv_{n,l,k} / 100$	

To define estimates, it is necessary to compare the meaningful characteristics of the resource types with generalized estimates proposed in Table 3.1, which, on our opinion, will require little effort to interpret the properties of the resource types and an appropriate qualitative assessment. For example, the first of the estimates in Table 3.1 "Availability and quality," when characterizing the type of "Health" resources of the "Social resources" category, will reflect the availability of a wellness facilities network, access to healthcare facilities, cost of treatment, availability of qualified doctors, etc.

Qualitative estimates of $q_{n,l,k,}$* will be attributed to all *n,l,k*-types of resources. In determining them, we will proceed from the state of resources, ways of their use, the expected state of current trends, etc. Other definitions of generalized estimates are possible too.

To define $q_{n,l,k'}$*, for example, a scale from 1 (lowest) to 10 (highest) can be used to reflect the status and current trends in resource use. Estimation of the expected state of the resource $qv_{n,l,k'}$* in Table 3.1 is the percentage decrease or increase of $q_{n,l,k'}$*. In addition to quantitative and qualitative estimates, we will consider (Table 3.2) the weight coefficients $w_{n,l}$ of the categories and $u_{n,l,k}$ of the types of resources.

Table 3.2 Weight coefficients and qualitative estimates of categories and types of resources

Categories	Types	Weight coefficients	Qualitative estimates	Weight coefficients
Social resources	Population (number)	$u_{n,1,1}$	$qav_{n,1,1}$	$w_{n,1}$
	Provision	$u_{n,1,2}$	$qav_{n,1,2}$	
	Housing provision	$u_{n,1,3}$	$qav_{n,1,3}$	
	Healthcare	$u_{n,1,4}$	$qav_{n,1,4}$	
	Security	$u_{n,1,5}$	$qav_{n,1,5}$	
	Education	$u_{n,1,6}$	$qav_{n,1,6}$	
	Communication and culture	$u_{n,1,7}$	$qav_{n,1,7}$	
	Civil activity	$u_{n,1,8}$	$qav_{n,1,8}$	
Natural resources	Land resources	$u_{n,2,1}$	$qav_{n,2,1}$	$w_{n,2}$
	Minerals	$u_{n,2,2}$	$qav_{n,2,2}$	
	Water resources	$u_{n,2,3}$	$qav_{n,2,3}$	
	Atmosphere	$u_{n,2,4}$	$qav_{n,2,4}$	
	Forest resources	$u_{n,2,8}$	$qav_{n,2,8}$	
	Fauna and flora	$u_{n,2,6}$	$qav_{n,2,6}$	
	Waste and hazardous chemicals	$u_{n,2,7}$	$qav_{n,2,7}$	
	Renovation activities financing	$u_{n,2,8}$	$qav_{n,2,8}$	
Economic resources	By types of economic activity	$u_{n,3,1}$ - $u_{n,3,21}$	$qav_{n,3,1}$ -$qav_{n,3,21}$	$w_{n,3}$
Infrastructural	Objects	$u_{n,4,1}$	$qav_{n,4,1}$	$w_{n,4}$
	Critical objects, hostage objects	$u_{n,4,2}$	$qav_{n,4,2}$	
Information resources	Development level	$u_{n,5,1}$	$qav_{n,5,1}$	
	Risk level	$u_{n,5,2}$	$qav_{n,5,2}$	$w_{n,5}$
	Information pressure	$u_{n,5,3}$	$qav_{n,5,3}$	
	Information war	$u_{n,5,4}$	$qav_{n,5,4}$	
Research and development resources	Additionally, they are allocated to the fields of research	$u_{n,6,1}$	$qav_{n,6,1}$	$w_{n,6}$
Governmental (regional) resources	Structure and activity	$u_{n,7,1}$	$qav_{n,7,1}$	$w_{n,7}$
	Budget funds	$u_{n,7,2}$	$qav_{n,7,2}$	

For weight coefficients $u_{n,l,k}$ and $w_{n,l}$ it is fulfilled—

$$\sum_k u_{n,l,k} = 1,\ 0 \le u_{n,l,k} \le 1,\ \sum_l w_{n,l}(t_j) = 1,\ 0 \le w_{n,l}(t_j) \le 1\ .$$

So, at the moment t_j we have constructed on the basis of (3.7) – (3.10) estimates of the *n,l,k*-type of resource taking into account the weight coefficients and qualitative estimates—

$$rpk_{n,l,k}(t_j) = rc_{n,l,k}(t_j) * u_{n,l,k}(t_j) * qav_{n,l,k}(t_j),\ qav_{n,l,k}(t_j) = \sum_{i=1}^{k(l)} qav_{n,l,k.i}(t_j)\ . \tag{3.11}$$

Based on (3.11) and the weights $w_{n,l}$ we can form weighted estimates for the categories *l* of region *n*

$$RDI_n(t_j) = \sum_{l=1}^{7} \sum_{k=1}^{K(l)} rpk_{n,l,k}(t_j) * w_{n,l}(t_j), \tag{3.12}$$

in sums *7* and *K(l)*, respectively are the number of categories and types of region *n* resources.

Thus, RDI_n forms a total estimation of the regional resources state.

We will call RDI_n the Regional Development Index.

On the RDI_n it is based the National Development Index *NDI*, defined, for example, as the minimum of RDI_n indices by all regions of the country *n* -

$NDI = \min_n RDI_n$.

Index RDI_n is focused on the calculation of aggregate or partial estimates of the use and creation of resources. The choice of a certain value of the index by the partial estimates allows defining and adopting balanced development areas, which should be done by the authorities of the region in agreement with the interests of society, business, and others. At the same time, the RDI_n index does not address the specific tasks identified in the 17 Sustainable Development Goals (SDGs), which require defining a certain set of resources and analyzing their status to find strategies for a dedicated SDG task solving.

In the case of the 10th goal of the SDG, "Reduced Inequalities" such a set is population that forms social and other resources and the tasks of finding strategies to use them to meet its needs. It also involves the interests and opportunities coordination of different social groups in developing and implementing strategies for solving the initial inequalities problem within the process of using and creating resources.

This requires an estimation of the current situation, its change, and the potential consequences of the developed and implemented strategies for the whole socio-ecological-economic system. For this purpose, it is proposed to apply the RDI_n index and its calculation, taking into account the features of the 10th SDG "Reduced Inequalities."

3.4 Inequality Index and Its Structure

The categories and types of resources (Table 3.2) were introduced above. We will proceed from the category of regional social resources, where first of all, problems of inequality arise and define the corresponding basic indicators (Table 3.3). But first, we need to identify the social groups whose interests and needs will be analyzed. We will assume only that the same social groups are distinguished within the country regions and that their union is consistent with the entire population. This gives the opportunity to apply diverse approaches to the social groups' selection, one of them is the level of regional resources using of which is estimated through indicators and summary indices (see below). That is, each of the social groups will be further described by a list of indicators reflecting the opportunities and results of using the resources available in the region.

Social groups can be defined, for example, based on M. Weber stratification [13], [14]. In it, classes are collections of people with the same life chances. In one class, capital is main, in another—qualifications. All the population was divided into owners; intellectuals (administrators and managers); small bourgeoisie (small entrepreneurs and merchants); and workers. Also belonging to a particular class is determined by the type of economic activity; the amount of income; type and level of education; place of residence; and type of leisure and recreation. Most of these criteria are further defined by regional resources and the level of their use in social groups.

Thus, the following basic indicators are introduced for social groups s (negative resources are highlighted (gray) and grouped separately).

This list of indicators certainly is not complete and indicates only the directions of its formation and refinement.

As a result, quantitatively the social group s status is be estimated by the following value of social resources (category 1):

$$Sr_{n,s,1}(t)=Sr_{no,s,1,1}(t)+\ldots+Sr_{n,s,1,8}(t). \tag{3.13}$$

Table 3.3 Category 1 "Social Resources"

	Basic indicators of resource types	Denotation
	1. Population (number)	$Sr_{n,s,1,1}(t)$
1	0–17 years	$pr_{n,s,1,1,1}(t)$
2	18–60 years	$pr_{n,s,1,1,2}(t)$
3	60 and older (pensioners)	$pr_{n,s,1,1,3}(t)$
4	Men	$pr_{n,s,1,1,4}(t)$
5	Women	$pr_{n,s,1,1,5}(t)$
6	Number of marriages	$pr_{n,s,1,1,6}(t)$
7	Number of divorces	$pr_{n,s,1,1,7}(t)$
8	Newborns	$pr_{n,s,1,1,8}(t)$
9	Immigrants	$pr_{n,s,1,1,9}(t)$
10	Emigrants	$dr_{n,s,1,1,10}(t)$
11	Dead	$dr_{n,s,1,1,11}(t)$
	$Sr_{n,s,1,1}(t)=pr_{n,s,1,1,1}(t)+...+pr_{n,s,1,1,9}(t)-(dr_{n,s,1,1,10}(t)+dr_{n,s,1,1,11}(t))$	
	2. Provision	$Sr_{n,s,1,2}(t)$
12	Number of men employed	$pr_{n,s,1,2,1}(t)$
13	Number of women employed	$pr_{n,s,1,2,2}(t)$
14	The amount of official wages	$pr_{n,s,1,2,3}(t)$
15	The amount of nutrition costs	$pr_{n,s,1,2,4}(t)$
16	Gross population savings	$pr_{n,s,1,2,5}(t)$
17	Money transfers to the population from abroad	$pr_{n,s,1,2,6}(t)$
18	The shadow wages amount	$dr_{n,s,1,2,7}(t)$
19	Population living below the poverty line	$dr_{n,s,1,2,8}(t)$
20	Number of people employed in unskilled or in work that does not match their qualifications	$dr_{n,s,1,2,9}(t)$
21	Number of registered unemployed men	$dr_{n,s,1,2,10}(t)$
22	Number of registered unemployed women	$dr_{n,s,1,2,11}(t)$
23	Number of unemployed with higher education	$dr_{n,s,1,2,12}(t)$
24	Number of diseases, accidents at manufacturing	$dr_{n,s,1,2,13}(t)$
25	Number of deaths at manufacturing	$dr_{n,s,1,2,14}(t)$
	$Sr_{n,s,1,2}(t)=pr_{n,s,1,2,1}(t)+...+pr_{n,s,1,2,6}(t)-(dr_{n,s,1,2,7}(t)+...+dr_{n,s,1,2,14}(t))$	
	3. Housing provision	$Sr_{n,s,1,3}(t)$
26	Number of households	$pr_{n,s,1,3,1}(t)$
27	Total living area	$pr_{n,s,1,3,2}(t)$
28	Volume of utilities	$pr_{n,s,1,3,3}(t)$

Continued

Table 3.3 *Continued*

	Basic indicators of resource types	Denotation
29	Population provided with quality drinking water	$dr_{n,s,1,3,4}(t)$
30	Number of households without electricity	$dr_{n,s,1,3,5}(t)$
31	Total area of emergency housing, housing without fixed assets	$dr_{n,s,1,3,6}(t)$
32	Expenditure on housing subsidies	$dr_{n,s,1,3,7}(t)$
	$Sr_{n,s,1,3}(t)=pr_{n,s,1,3,1}(t)+pr_{n,s,1,3,2}(t)+pr_{n,s,1,3,3}(t)-(dr_{n,s,1,3,4}(t)+...+dr_{n,s,1,3,7}(t))$	
	4. Health care	$Sr_{n,s,1,4}(t)$
33	Number of health facilities	$pr_{n,s,1,4,1}(t)$
34	Number of health facilities visitors	$pr_{n,s,1,4,2}(t)$
35	Number of children immunized against infectious diseases	$pr_{n,s,1,4,3}(t)$
36	Number of people with access to primary care	$pr_{n,s,1,4,4}(t)$
37	Number of doctors	$pr_{n,s,1,4,5}(t)$
38	Number of diseases	$pr_{n,s,1,4,6}(t)$
39	Number of diseases due to environmental pollution	$pr_{n,s,1,4,7}(t)$
40	The amount of health care expenditures	$pr_{n,s,1,4,8}(t)$
41	The amount of doctors' payroll	$pr_{n,s,1,4,9}(t)$
42	Number of hospital beds	$pr_{n,s,1,4,10}(t)$
43	Number of referrals to medical institutions	$pr_{n,s,1,4,11}(t)$
44	Number of cancer patients	$dr_{n,s,1,4,12}(t)$
45	Number of patients with cardiovascular diseases	$dr_{n,s,1,4,13}(t)$
46	The number of patients with dangerous diseases (HIV/AIDS, tuberculosis, etc.)	$dr_{n,s,1,4,14}(t)$
47	The number of patients with mental pathology	$dr_{n,s,1,4,15}(t)$
48	Number of deaths under 5 years	$dr_{n,s,1,4,16}(t)$
49	The number of drug addicts	$dr_{n,s,1,4,17}(t)$
50	The number of patients with alcoholism	$dr_{n,s,1,4,18}(t)$
51	Volume of alcohol consumption	$dr_{n,s,1,4,19}(t)$
	$Sr_{n,s,1,4}(t)=pr_{n,s,1,4,1}(t)+...+pr_{n,s,1,4,11}(t)-(dr_{n,s,1,4,12}(t)+...+dr_{n,s,1,4,19}(t))$	
	5. Security	$Sr_{n,s,1,5}(t)$
52	Number of law enforcement officers	$pr_{n,s,1,5,1}(t)$
53	Number of courts	$pr_{n,s,1,5,2}(t)$
54	Number of convicts	$dr_{n,s,1,5,3}(t)$
55	Number of minor crimes	$dr_{n,s,1,5,4}(t)$
56	Number of fraudulent crimes	$dr_{n,s,1,5,5}(t)$
57	Number of cybercrimes	$dr_{n,s,1,5,6}t)$

Continued

Table 3.3 ***Continued***

	Basic indicators of resource types	**Denotation**
58	Number of violent crimes	$dr_{n,s,1,5,7}(t)$
59	Number of criminal groups	$dr_{n,s,1,5,8}(t)$
60	Number of premeditated murders	$dr_{n,s,1,5,9}(t)$
61	Number of appeals to courts	$dr_{n,s,1,5,10}(t)$
62	Economic losses of the population due to natural disasters or crisis phenomena	$dr_{n,s,1,5,11}(t)$
63	Number of deaths due to natural disasters or crises phenomena	$dr_{n,s,1,5,12}(t)$
	$Sr_{n,s,1,5}(t)=pr_{n,s,1,5,1}(t)+pr_{n,s,1,5,2}(t)-(dr_{n,s,1,5,3}(t)+\ldots+dr_{n,s,1,5,12}(t))$	
	6. Education	$Sr_{n,s,1,6}(t)$
64	Number of children in pre-school institutions	$pr_{n,s,1,6,1}(t)$
65	Number of school age children	$pr_{n,s,1,6,2}(t)$
66	Number of people attending school	$pr_{n,s,1,6,3}(t)$
67	Number of school graduates	$pr_{n,s,1,6,4}(t)$
68	Number of adults with secondary education	$pr_{n,s,1,6,5}(t)$
69	Number of pre-school teachers	$pr_{n,s,1,6,6}(t)$
70	Number of schoolteachers	$pr_{n,s,1,6,7}(t)$
71	Number of higher education institutions	$pr_{n,s,1,6,8}(t)$
72	Number of universities with global rankings	$pr_{n,s,1,6,9}(t)$
73	Number of teachers at higher education	$pr_{n,s,1,6,10}(t)$
74	Number of students at higher education	$pr_{n,s,1,6,11}(t)$
75	Number of international students	$pr_{n,s,1,6,12}(t)$
76	Number of scientific publications	$pr_{n,s,1,6,13}(t)$
77	Number of e-courses used	$pr_{n,s,1,6,14}(t)$
78	Number of e-courses users	$pr_{n,s,1,6,15}(t)$
79	Number of universities offering e-education diplomas	$pr_{n,s,1,6,16}(t)$
80	Expenditure on pre-school education	$pr_{n,s,1,6,17}(t)$
81	Volume of private and public investment in pre-school education, including tuition fees	$pr_{n,s,1,6,18}(t)$
82	Expenditure on secondary education	$pr_{n,s,1,6,19}(t)$
83	Volume of private and public investment in secondary education, including tuition fees	$pr_{n,s,1,6,20}(t)$
84	Higher education expenditures	$pr_{n,s,1,6,21}(t)$
85	Volume of private and public investment in higher education, including tuition fees	$pr_{n,s,1,6,22}(t)$

Continued

Table 3.3 *Continued*

	Basic indicators of resource types	Denotation
86	Number of men with higher education	$pr_{n,s,1,6,23}(t)$
87	Number of women with higher education	$pr_{n,s,1,6,24}(t)$
	$Sr_{n,s,1,6}(t)=pr_{n,s,1,6,1}(t)+...+pr_{n,s,1,6,24}(t))$	
	7. Communication and culture	$Sr_{n,s,1,7}(t)$
88	Number of computers for personal use	$pr_{n,s,1,7,1}(t)$
89	Number of Internet users	$pr_{n,s,1,7,2}(t)$
90	Number of users of social networks	$pr_{n,s,1,7,3}(t)$
91	Volume of communication services	$pr_{n,s,1,7,4}(t)$
92	Number of theaters, cinemas, museums	$pr_{n,s,1,7,5}(t)$
93	Number of visits to theaters, movies, museums	$pr_{n,s,1,7,6}(t)$
94	Expenditure on culture, media, sports, etc.	$pr_{n,s,1,7,7}(t)$
	$Sr_{n,s,1,7}(t)=pr_{n,s,1,7,1}(t)+...+pr_{n,s,1,7,7}(t))$	
	8. Civil activity	$Sr_{n,s,1,8}(t)$
95	Number of major confessions	$pr_{n,s,1,8,1}(t)$
	The number of people in major confessions	
96	confession 1	$pr_{n,s,1,8,2}(t)$
97	confession 2	$pr_{n,s,1,8,3}(t)$
98	confession 3	$pr_{n,s,1,8,4}(t)$
99	Number of participants in public consultations, discussions, etc.	$pr_{n,s,1,8,5}(t)$
100	Number of civil patrol members	$pr_{n,s,1,8,6}(t)$
101	The number of voters who participated in state-level elections	$pr_{n,s,1,8,7}(t)$
102	The number of voters who participated in the regional elections	$pr_{n,s,1,8,8}(t)$
103	Number of strikes, demonstrations, etc.	$dr_{n,s,1,8,9}(t)$
104	Number of participants in strikes, demonstrations, etc.	$dr_{n,s,1,8,10}(t)$
105	The population that gave bribes	$dr_{n,s,1,8,11}(t)$
106	Number of internal conflicts	$dr_{n,s,1,8,12}(t)$
107	Number of participants in internal conflicts	$dr_{n,s,1,8,13}(t)$
108	Number of separatism manifestations	$dr_{n,s,1,8,14}(t)$
109	Number of suicides	$dr_{n,s,1,8,15}(t)$
	$Sr_{n,s,1,8}(t)=pr_{n,s,1,8,1}(t)+...+pr_{n,s,1,8,8}(t))$ - $(dr_{n,s,1,8,9}(t)+...+dr_{n,s,1,8,15}(t))$	

Based on the estimates of Tables 3.2 and 3.3, together with (3.13), we introduce quantitative similar estimates of other social group *s* resources:

- natural (category 2)—$Sr_{n,s,2}(t) = Sr_{n,s,2,1}(t)+...+Sr_{n,s,2,8}(t);$
- economic—$Sr_{n,s,3}(t) = Sr_{n,s,3,1}(t) + ... + Sr_{n,s,3,21}(t);$
- infrastructural—$Sr_{n,s,4}(t) = Sr_{n,s,4,1}(t) + Sr_{n,s,4,2}(t);$
- information—$Sr_{n,s,5}(t) = Sr_{n,s,5,1}(t) + ... + Sr_{n,s,5,4}(t);$
- research and development—$Sr_{n,s,6}(t) = Sr_{n,s,6,1}(t) + ... + Sr_{n,s,6,x}(t),$ are allocated to the research fields $x;$
- governmental—$Sr_{n,s,7}(t) = Sr_{n,s7,1}(t) + Sr_{n,s,7,2}(t).$

We have an estimate of all social group *s* resources—

$$S_{n,s}(t)=Sr_{n,s,1}(t)+Sr_{n,s,2}(t)+Sr_{n,s,3}(t)+Sr_{n,s,4}(t)+Sr_{n,s,5}(t)+Sr_{n,s,6}(t)+Sr_{n,s,7}(t). \quad (3.14)$$

which reflects the quantitative estimate of this group's ability to create new resources based on their volume. Comparison of this estimate gives a priori estimate of inequality level in the use of resources by social groups in one region and for the same group *s* in different regions.

To consider the properties of resources, the results of their use, in addition to quantitative estimates (3.14) we will consider their qualitative ones. To reduce the amount of information in Table 3.1 generalized estimates were introduced that focused on all types of resources. In our case, we will introduce different estimates for all types of category 1 "Social resources."

As introduced quantitative, qualitative estimates are not the final result, but only indicate the directions of development, the first of them $q_{n,s,1,1,1}(t)$, for example, indicates the first qualitative estimate of a category 1 type 1 resource of social group *s* of the region *n* at time *t*.

The estimates $q_{n,s,1,k,i}(t)$ will also be defined by a 10-point scale, but unlike the previous ones (Table 3.1) with values ranging from 0 (absent) to 9 (maximum). The expected state of the resource $qv_{n,s,l,k,i}(t)$ in Table 3.4 also indicates the percentage changes $q_{n,s,l,k,i}(t)$.

By ranking of the estimates $Sq_{n,s,1}(t)$ by *n* or *s*, we obtain, respectively, the level of group *s* social resources in all regions *n*, or all groups *s* in region *n*. The rankings will reflect the "qualitative" inequality of groups by social resources, in particular because of the relation of individual estimates to their sum.

Let's define similar (3.11) cumulative quantitative and qualitative weighted estimates by resources types k and category 1 "Social resources" of the group *s*—

$$SQ_{n,s,1,k}(t)=Sr_{n,s,1,k}(t)*qav_{n,s,1,k}(t)*u_{n,s,1,k}(t),$$

Table 3.4 Category 1 "Social resources"—qualitative estimates for social group *s*

	1. Population (number)	Current trends	The expected state of current trends (± %)
1	Necessity, pressure (for example, immigration) from other regions, countries, resource categories	$q_{n,s,1,1,1}(t)$	$qv_{n,s,1,1,1}(t)$
2	Impact on other resources, regions, countries	$q_{n,s,1,1,2}(t)$	$qv_{n,s,1,1,2}(t)$
3	Average age	$q_{n,s,1,1,3}(t)$	$qv_{n,s,1,1,3}(t)$
4	Qualification	$q_{n,s,1,1,4}(t)$	$qv_{n,s,1,1,4}(t)$
5	The emigration level, including labor	$q_{n,s,1,1,5}(t)$	$qv_{n,s,1,1,5}(t)$
6	The immigration level	$q_{n,s,1,1,6}(t)$	$qv_{n,s,1,1,6}(t)$
7	Self-improvement	$q_{n,s,1,1,7}(t)$	$qv_{n,s,1,1,7}(t)$
8	External assistance	$q_{n,s,1,1,8}(t)$	$qv_{n,s,1,1,8}(t)$
9	Outcome of external assistance	$q_{n,s,1,1,9}(t)$	$qv_{n,s,1,1,9}(t)$
10	The positive change potential	$q_{n,s,1,1,10}(t)$	$qv_{n,s,1,1,10}(t)$
	Resource category 1 type 1 (arithmetic average, normalized by maximum scale value)	$q_{n,s,1,1}(t)=\dfrac{\sum_{i=1}^{10} q_{n,s,1,1,i}(t)}{9*10}$	$qv_{n,s,1,1}(t)=\dfrac{\sum_{i=1}^{10} qv_{n,s,1,1,i}(t)}{9*10}$
		$qav_{n,s,1,1}(t)=(q_{n,s,1,1}(t)+qv_{n,s,1,1}(t)/100)/2$	
	2. Provision		
11	The level of provision	$q_{n,s,1,2,1}(t)$	$qv_{n,s,1,2,1}(t)$
12	Need, pressure	$q_{n,s,1,2,2}(t)$	$qv_{n,s,1,2,2}(t)$
13	Impact on other resources	$q_{n,s,1,2,3}(t)$	$qv_{n,s,1,2,3}(t)$
14	Working day duration	$q_{n,s,1,2,4}(t)$	$qv_{n,s,1,2,4}(t)$
15	Time for rest, leisure	$q_{n,s,1,2,5}(t)$	$qv_{n,s,1,2,5}(t)$
16	The level of economic demarcation of the population	$q_{n,s,1,2,6}(t)$	$qv_{n,s,1,2,6}(t)$
17	Perceptions of the provision level by population	$q_{n,s,1,2,7}(t)$	$qv_{n,s,1,2,7}(t)$
18	Self-estimation of the population quality of life	$q_{n,s,1,2,8}(t)$	$qv_{n,s,1,2,8}(t)$
19	Poverty risk	$q_{n,s,1,2,9}(t)$	$qv_{n,s,1,2,9}(t)$
20	The support level of unemployed	$q_{n,s,1,2,10}(t)$	$qv_{n,s,1,2,10}(t)$
21	The positive change potential	$q_{n,s,1,2,11}(t)$	$qv_{n,s,1,2,11}(t)$
	Resource category 1 type 2	$q_{n,s,1,2}(t)=\dfrac{\sum_{i=1}^{11} q_{n,s,1,2,i}(t)}{9*11}$	$qv_{n,s,1,2}(t)=\dfrac{\sum_{i=1}^{11} qv_{n,s,1,2,i}(t)}{9*11}$
		$qav_{n,s,1,2}(t)=(q_{n,s,1,2}(t)+qv_{n,s,1,2}(t)/100)/2$	

Continued

Table 3.4 *Continued*

	3. Housing provision		
22	Housing quality level	$q_{n,s,1,3,1}(t)$	$qv_{n,s,1,3,1}(t)$
23	Provision level of basic utilities	$q_{n,s,1,3,2}(t)$	$qv_{n,s,1,3,2}(t)$
24	Attitude to the basic services provision	$q_{n,s,1,3,3}(t)$	$qv_{n,s,1,3,3}(t)$
25	The level of contamination of residential areas	$q_{n,s,1,3,4}(t)$	$qv_{n,s,1,3,4}(t)$
26	Attitude to the residential areas' pollution	$q_{n,s,1,3,5}(t)$	$qv_{n,s,1,3,5}(t)$
27	The utility bills non-payment level	$q_{n,s,1,3,6}(t)$	$qv_{n,s,1,3,6}(t)$
28	Population energy saving measures	$q_{n,s,1,3,7}(t)$	$qv_{n,s,1,3,7}(t)$
29	Authorities measures to reduce tariffs	$q_{n,s,1,3,8}(t)$	$qv_{n,s,1,3,8}(t)$
30	Authorities energy saving measures	$q_{n,s,1,3,9}(t)$	$qv_{n,s,1,3,9}(t)$
31	The positive change potential	$q_{n,s,1,3,10}(t)$	$qv_{n,s,1,3,10}(t)$
	Resource category 1 type 3	$q_{n,s,1,3}(t) = \frac{\sum_{i=1}^{10} q_{n,s,1,3,i}(t)}{9*10}$	$qv_{n,s,1,3} = \frac{\sum_{i=1}^{10} qv_{n,s,1,3,i}(t)}{9*10}$
		$qav_{n,s,1,3}(t) = (q_{n,s,1,3}(t) + qv_{n,s,1,3}(t)/100)/2$	
	4. Health care		
32	Stability of the health care system and its development	$q_{n,s,1,4,1}(t)$	$qv_{n,s,1,4,1}(t)$
33	Need, pressure	$q_{n,s,1,4,2}(t)$	$qv_{n,s,1,4,2}(t)$
34	Compliance of medical institutions to the world level	$q_{n,s,1,4,3}(t)$	$qv_{n,s,1,4,3}(t)$
35	Condition of premises and equipment	$q_{n,s,1,4,4}(t)$	$qv_{n,s,1,4,4}(t)$
36	Condition and level of basic provision	$q_{n,s,1,4,5}(t)$	$qv_{n,s,1,4,5}(t)$
37	Personnel qualification	$q_{n,s,1,4,6}(t)$	$qv_{n,s,1,4,6}(t)$
38	Personnel sufficiency of all qualifications	$q_{n,s,1,4,7}(t)$	$qv_{n,s,1,4,7}(t)$
39	Personnel wages	$q_{n,s,1,4,8}(t)$	$qv_{n,s,1,4,8}(t)$
40	Stability of medicines supply	$q_{n,s,1,4,9}(t)$	$qv_{n,s,1,4,9}(t)$
41	The level of ambulance care, including cost	$q_{n,s,1,1410}(t)$	$qv_{n,s,1,1410}(t)$
42	The level of hospital care, including cost	$q_{n,s,1,4,11}(t)$	$qv_{n,s,1,4,11}(t)$

Continued

Table 3.4 Continued

43	Need for treatment in other regions, including abroad	$q_{n,s,1,4,12}(t)$	$qv_{n,s,1,4,12}(t)$
44	The development level of health facilities	$q_{n,s,1,4,13}(t)$	$qv_{n,s,1,4,13}(t)$
45	Level of service in health facilities, including cost	$q_{n,s,1,4,14}(t)$	$qv_{n,s,1,4,14}(t)$
46	The attitude level to the healthy lifestyle	$q_{n,s,1,4,15}(t)$	$qv_{n,s,1,4,15}(t)$
47	Health self-esteem level	$q_{n,s,1,4,16}(t)$	$qv_{n,s,1,4,16}(t)$
48	Drugs attitude level	$q_{n,s,1,4,17}(t)$	$qv_{n,s,1,4,17}(t)$
49	Alcohol attitude level	$q_{n,s,1,4,18}(t)$	$qv_{n,s,1,4,18}(t)$
50	Obesity level	$q_{n,s,1,4,19}(t)$	$qv_{n,s,1,4,19}(t)$
	Resource category 1 type 4	$q_{n,s,1,4}(t)=\frac{\sum_{i=1}^{19} q_{n,s,1,4,i}(t)}{9*19}$	$qv_{n,s,1,4}(t)=\frac{\sum_{i=1}^{19} qv_{n,s,1,4,i}(t)}{9*19}$
		$qav_{n,s,1,4}(t)=(q_{n,s,1,4}(t)+qv_{n,s,1,4}(t)/100)/2$	
	5. Security		
51	Security level	$q_{n,s,1,5,1}(t)$	$qv_{n,s,1,5,1}(t)$
52	Pressure from other regions, countries, resources categories	$q_{n,s,1,5,2}(t)$	$qv_{n,s,1,5,2}(t)$
53	Stability in the society	$q_{n,s,1,5,3}(t)$	$qv_{n,s,1,5,3}(t)$
54	Self-assessment of the individual security level	$q_{n,s,1,5,4}(t)$	$qv_{n,s,1,5,4}(t)$
55	Respect for others and their needs, norms of social behavior	$q_{n,s,1,5,5}(t)$	$qv_{n,s,1,5,5}(t)$
56	Attitude to the others property, to the property of collective use	$q_{n,s,1,5,6}(t)$	$qv_{n,s,1,5,6}(t)$
57	Attitude to the security and crime	$q_{n,s,1,5,7}(t)$	$qv_{n,s,1,5,7}(t)$
58	Attitude to the region situation, satisfaction with it	$q_{n,s,1,5,8}(t)$	$qv_{n,s,1,5,8}(t)$
	Resource category 1 type 5	$q_{n,s,1,5}(t)=\frac{\sum_{i=1}^{8} q_{n,s,1,5,i}(t)}{9*8}$	$qv_{n,s,1,5}(t)=\frac{\sum_{i=1}^{8} qv_{n,s,1,5,i}(t)}{9*8}$
		$qav_{n,s,1,5}(t)=(q_{n,s,1,5}(t)+qv_{n,s,1,5}(t)/100)/2$	
	6. Education		
59	The level of institutions and education quality	$q_{n,s,1,6,1}(t)$	$qv_{n,s,1,6,1}(t)$

Continued

Table 3.4 *Continued*

60	The level of institutions and higher education quality by international rankings	$q_{n,s,1,6,2}(t)$	$qv_{n,s,1,6,2}(t)$
61	The level of higher education cost	$q_{n,s,1,6,3}(t)$	$qv_{n,s,1,6,3}(t)$
62	Secondary education accessibility	$q_{n,s,1,6,4}(t)$	$qv_{n,s,1,6,4}(t)$
63	Higher education accessibility	$q_{n,s,1,6,5}(t)$	$qv_{n,s,1,6,5}(t)$
64	Need, pressure	$q_{n,s,1,6,6}(t)$	$qv_{n,s,1,6,6}(t)$
65	Time spent on education	$q_{n,s,1,6,7}(t)$	$qv_{n,s,1,6,7}(t)$
66	The importance of education for success in the job market, own business	$q_{n,s,1,6,8}(t)$	$qv_{n,s,1,6,8}(t)$
67	Interest in higher education	$q_{n,s,1,6,9}(t)$	$qv_{n,s,1,6,9}(t)$
68	The level of information technologies use in education	$q_{n,s,1,6,10}(t)$	$qv_{n,s,1,6,10}(t)$
69	The attitude level of innovative technologies study	$q_{n,s,1,6,11}(t)$	$qv_{n,s,1,6,11}(t)$
70	The level of population readiness for the continuing education	$q_{n,s,1,6,12}(t)$	$qv_{n,s,1,6,12}(t)$
71	The population readiness level for the e-education	$q_{n,s,1,6,13}(t)$	$qv_{n,s,1,6,13}(t)$
	Resource category 1 type 6	$q_{n,s,1,6}(t) = \frac{\sum_{i=1}^{13} q_{n,s,1,6,i}(t)}{9*13}$	$qv(t)_{n,s,1,6} = \frac{\sum_{i=1}^{13} qv(t)_{n,s,1,6,i}}{9*13}$
		$qav(t)_{n,s,1,6} = (q(t)_{n,s,1,6} + qv(t)_{n,s,1,6} / 100) / 2$	
	7. Communication and culture		
72	Need, pressure	$q_{n,s,1,7,1}(t)$	$qv_{n,s,1,7,1}(t)$
73	The level of information technologies uses in current life and own business	$q_{n,s,1,7,2}(t)$	$qv_{n,s,1,7,2}(t)$
74	Protection of personal rights and data level	$q_{n,s,1,7,3}(t)$	$qv_{n,s,1,7,3}(t)$
75	Internet addiction level	$q_{n,s,1,7,4}(t)$	$qv_{n,s,1,7,4}(t)$
76	Tendency to group life, mutual respect and assistance	$q_{n,s,1,7,5}(t)$	$qv_{n,s,1,7,5}(t)$
77	Culture, attitude to work, etc.	$q_{n,s,1,7,6}(t)$	$qv_{n,s,1,7,6}(t)$
78	Individualism	$q_{n,s,1,7,7}(t)$	$qv_{n,s,1,7,7}(t)$
79	Immigration tolerance	$q_{n,s,1,7,8}(t)$	$qv_{n,s,1,7,8}(t)$
80	Minority discrimination	$q_{n,s,1,7,9}(t)$	$qv_{n,s,1,7,9}(t)$

Continued

Table 3.4 Continued

81	The positive change potential Resource category 1 type 7	$q_{n,s,1,7,10}(t)$ $q_{n,s,1,7}(t)=\frac{\sum_{i=1}^{10} q(t)_{n,s,1,7,i}}{9*10}$	$qv_{n,s,1,7,10}(t)$ $qv_{n,s,1,7}(t)=\frac{\sum_{i=1}^{10} qv_{n,s,1,7,i}(t)}{9*10}$
		$qav_{n,s,1,7}(t)=(q_{n,s,1,7}(t)+qv_{n,s,1,7}(t)/100)/2$	
	7. Civil activity		
82	Civil activity level	$q_{n,s,1,8,1}(t)$	$qv_{n,s,1,8,1}(t)$
83	Pressure from other regions, countries, resources categories	$q_{n,s,1,8,2}(t)$	$qv_{n,s,1,8,2}(t)$
84	Impact on the population of other regions and countries	$q_{n,s,1,8,3}(t)$	$qv_{n,s,1,8,3}(t)$
85	Religiosity of society	$q_{n,s,1,8,4}(t)$	$qv_{n,s,1,8,4}(t)$
86	Religious tolerance	$q_{n,s,1,8,5}(t)$	$qv_{n,s,1,8,5}(t)$
87	The social tension level	$q_{n,s,1,8,6}(t)$	$qv_{n,s,1,8,6}(t)$
88	Attitude to authorities, trust and support of their activities	$q_{n,s,1,8,7}(t)$	$qv_{n,s,1,8,7}(t)$
89	Propensity for acts of disobedience, criminal and terrorist acts and attitude to them	$q_{n,s,1,8,8}(t)$	$qv_{n,s,1,8,8}(t)$
90	The level of socio-political demarcation in society	$q_{n,s,1,8,9}(t)$	$qv_{n,s,1,8,9}(t)$
91	The probability of socio-political conflicts, violent demonstrations	$q_{n,s,1,8,10}(t)$	$qv_{n,s,1,8,10}(t)$
92	The conflict danger level	$q_{n,s,1,8,11}(t)$	$qv_{n,s,1,8,11}(t)$
93	The probability of weapons using in a conflict	$q_{n,s,1,8,12}(t)$	$qv_{n,s,1,8,12}(t)$
94	Probability of conflict spread	$q_{n,s,1,8,13}(t)$	$qv_{n,s,1,8,13}(t)$
95	Probability of conflict damping	$q_{n,s,1,8,14}(t)$	$qv_{n,s,1,8,14}(t)$
96	Internal forces conflicts support	$q_{n,s,1,8,15}(t)$	$qv_{n,s,1,8,15}(t)$
97	External conflict support	$q_{n,s,1,5,16}(t)$	$qv_{n,s,1,5,16}(t)$
98	The positive change potential Resource category 1 type 8	$q_{n,s,1,8,17}(t)$ $q_{n,s,1,8}(t)=\frac{\sum_{i=1}^{17} q_{n,s,1,8,i}(t)}{9*17}$	$qv_{n,s,1,8,17}(t)$ $qv_{n,s,1,8}(t)=\frac{\sum_{i=1}^{17} qv_{n,s,1,8,i}(t)}{9*17}$
		$qav_{n,s,1,8}(t)=(q_{n,s,1,8}(t)+qv_{n,s,1,8}(t)/100)/2$	
	Total category 1 estimation of region *n* group *s* "Social resources"	$Sq_{n,s,1}(t)=qav_{n,s,1,1}(t)+...+qav_{n,s,1,8}(t)$	

in the general case we have—

$$SQ_{n,s,l,k}(t) = Sr_{n,s,l,k}(t) * qav_{n,s,l,k}(t) * u_{n,s,l,k}(t).$$

The resource category is thus estimated as—

$$SQ_{n,s,l}(t) = \sum_{k=1}^{K(l)} SQ_{n,s,l,k}(t) * w_{n,s,l}(t) ,$$

for all 7 resource categories of group *s*. Estimates for all other categories are determined by analogy with social resources, we have—

$$SQ_{n,s}(t) = \sum_{l=1}^{7} SQ_{n,s,l}(t) = \sum_{l=1}^{7} \sum_{k=1}^{K(l)} SQ_{n,s,l,k}(t) * w_{n,s,l}(t) . \tag{3.15}$$

The value (3.15) thus coincides with the regional development index RDI_n (3.12), which uses detailed qualitative estimates of resource types. Ratings based on (3.15), as well as their qualitative components, also allow estimating corresponding resource status by region and social group.

Further, for convenience, we will also define generalized notations of the form $SQ_{n,s,l,k}(t)$ by the names of the resource categories (Table 3.5).

The relations in the second column of the table form partial estimates, by which comparisons by region or social group allow to define a comparative estimate of the state of group *s*. That is, after comparing this group with other groups, it is possible to define its partial inequality by resource category, which allows accepting these relations as subindices of comparative state of these group resources.

A full inequality estimate is defined by the aggregate of all resources. There are various options, including:

$$SP_{n,s}(t) = SN_{n,s}(t) + SE_{n,s}(t) + SS_{n,s}(t) + SI_{n,s}(t) + SD_{n,s}(t) + SG_{n,s}(t)$$

Table 3.5 Resource categories and subindices inequality of social group s

Resource categories	Relative estimates
$SQ_{n,s,1}(t) = SC_{n,s}(t)$—social	-
$SQ_{n,s,2}(t) = NT_{n,s}(t)$—natural	$SN_{n,s}(t) = SC_{n,s}(t)/NT_{n,s}(t)$
$SQ_{n,s,3}(t) = EC_{n,s}(t)$—economic	$SE_{n,s}(t) = SC_{n,s}(t)/EC_{n,s}(t)$
$SQ_{n,s,4}(t) = IS_{n,s}(t)$—infrastructural	$SS_{n,s}(t) = SC_{n,s}(t)/IS_{n,s}(t)$
$SQ_{n,s,5}(t) = IT_{n,s}(t)$—information	$SI_{n,s}(t) = SC_{n,s}(t)/IT_{n,s}(t)$
$SQ_{n,s,6}(t) = RD_{n,s}(t)$—research and development	$SD_{n,s}(t) = SC_{n,s}(t)/RD_{n,s}(t)$
$SQ_{n,s,7}(t) = GV_{n,s}(t)$—governmental	$SG_{n,s}(t) = SC_{n,s}(t)/GV_{n,s}(t)$

or

$$SR_{n,s}(t) = SC_{n,s}(t) / (SC_{n,s}(t) + SN_{n,s}(t) + SE_{n,s}(t) + SS_{n,s}(t) + SI_{n,s}(t) + SD_{n,s}(t) + SG_{n,s}(t)) =$$
$$= SC_{n,s}(t) / SQ_{n,s}(t),$$

where the group's social resources are treated as a part of their sum.

Then the minimum value of $SR_{n,s}(t)$ by all social groups s of one region n or by all regions n for one social group s will accordingly reflect the worst estimate of their inequality in terms of available resources, that is, define the largest inequality of group status—

$$MS_n(t) = \min_s SR_{n,s}(t) \text{ та } MN_s(t) - \min_n SR_{n,s}(t)$$

and a minimum by s and n will indicate the worst by level social group s in all regions n of the country—

$$II(t) = \min_{n,s} SR_{n,s}(t) . \tag{3.16}$$

We will define (3.16) as the Inequality Index of a country for all its social groups s and regions n.

By analogy with (3.16), an index for different countries is being constructed, it is sufficient to define resources as a share of their global volume.

3.5 Inequality Index and Directions of Its Use

The proposed Inequality Index $II(t)$ and its components make it possible to define the status of regions and their social groups, based on comparative information on available resources, that is, on objective quantitative estimate of what the social group of the region has and what is the basis for its further activity and limit of its potential outcomes. In addition, these estimations are enhanced qualitatively by possibilities and outcomes of resource use by social group and include the coefficients of importance for different regional resources.

The matter is that just from the available resources you can achieve something that, in addition to resources, depends on the desire, qualification, experience, work, the original goal, at least, on luck, etc. For governing bodies, the most important is also the potential for change, which can be objectively defined at the expense of all available resources, ways of using them, as well as their correlation with other regions and countries, including the inequality problem solving.

The state of resources allows to define the index $II_{n,s}(t)$ by groups s and regions n, it also includes the quantitative and qualitative estimates that reduce its value the most. There are these values that substantiate the areas of activity of the region, the choice, and ways of using critical resources that should be directed to increase the estimates and $II_{n,s}(t)$.

For this purpose, for the smallest or for the other selected by groups s and regions n index $II_{n,s}(t)$, after the transition to its critical minimum quantitative and qualitative estimates, first the possibilities for their increase must be found and defined. These possibilities may be limited by age, economic and other inequalities, which, being objective, are traditional, and we don't overcome them. They will require significant costs, time to change the population mentality, norms, habits, etc. The index also forms measures to combat inequality based on the critical estimates obtained from the calculation and comparison of the indices $II_{n,s}(t)$.

Thus, the index $II_{n,s}(t)$ formally defines the basics for developing measures to improve these estimates that can be put into practice. The next step is to select among them the most effective measures, considering the results of the use of resources available in the regions and social groups, as well as ways of behavior in relation to these resources or the social life norms.

Taken together, these actions form effective inequality reducing strategies that can vary significantly across social groups and regions and the potential outcomes of their implementation. For example,, Ukraine has different natural, industrial and other resources, which does not allow to find a common strategy for their development, overcoming mental, economic and other components of inequality, but require different efforts of the regional authorities and territorial communities to do this.

3.6 Conclusion

The constructed $II_{n,s}(t)$ is being tested today on practical calculations, but more interesting is its extension to other SDG tasks and the creation of a common methodology for the formal estimation of the problem and its components as a whole, and the definition of appropriate clear measures for its implementation.

Today, the index creates good prospects for success in this work. An important area of research is to reduce data, including replacing them with available UN data and using them for expert analysis. At the same time, this information needs to be complemented by countries' regional data, not just using national data. This will allow finding real effective strategies for regional and further national development and for reducing inequalities.

References

1. World Commission on Environment and Development (1987). Our Common Future. Oxford: Oxford University Press. p. 27. ISBN 019282080X. https://sustainabledevelopment.un.org/content/documents/5987our-common-future.pdf
2. Gazdar K. (1998). Germany's Balanced Development: The Real Wealth of a Nation, 1998. - https://books.google.com.ua/books?id=SQb7pkSl0asC&pg=SA84&lpg=SA84&dq=balanced+development&source=bl&ots=SvnKo1WStv&sig=Ubsnf-wuHOWIS_kleyHEXz8ZrF-w&hl=ru&sa=X&ved=0ahUKEwic3rv-rpjYAhVJL1AKHXIICBw4ChDoAQgmMAA#v= onepage&q=balanced%20development&f=false.
3. United Nations (2013). Global Sustainable Development Report. Executive Summary: Building the Common Future We Want. New York: United Nations Department of Economic and Social Affairs, Division for Sustainable Development. 2013. - http://sustainabledevelopment.un.org/globalsdreport.
4. IISD (1999). Indicators for Sustainable Development: Theory, Method, Applications: a Report to the Balaton Group. International Institute for Sustainable Development, 1999.
5. United Nations (2019). Goal 10: Reduce inequality within and among countries. https://www.un.org/sustainabledevelopment/inequality/
6. Gorda, S. and Polumiienko, S. (2016). "Some methods of indicative assessment of national development processes", The mechanism of the economy regulation, 2, pp. 83-95, 2016 (in Ukrainian).
7. Polumiienko, and Rybakov, S. L. (2015). Indicative analysis of the sustainability of technological development, Kyiv, Logos, 2015 (in Ukrainian).
8. Human Development Report (2019). Beyond income, beyond averages, beyond today: Inequalities in human development in the 21st century. -http://hdr.undp.org/sites/default/files/hdr_2019_overview_-_russian.pdf.
9. Government of Ukraine (2020). Balanced Development Plan. - http://english.seoul.go.kr/policy-information/urban-planning/balanced-development-plan/.
10. Polumiienko, S.; O. Stryzhak, O. Trofymchuk, (2016). "Resource-ontological approach to the assessment of the national development level", Mathematical modeling in economics, 7 (3-4), pp. 7–26, 2016 (in Ukrainian).
11. Danilov, V. (2002). Lectures on game theory, oscow, Russian Economic School, 2002 (in Russian).

12. Aumann, R.J. L.S. Shapley, (1974). Values of Non-Atomic Games, Princeton, Princeton Univ. Press, 1974.
13. Weber, M. (2009). The Theory of Social and Economic Organization. Simon and Schuster, 2009. - https://books.google.com.ua/books?id=G3TYBu6-4G0C&dq=inauthor:%22Max+Weber%22&hl=ru&source=gbs_navlinks_s.
14. Dvorets'ka, G. (2002). Sociology. Educational manual, Kyiv, KNEU, 2002. - https://buklib.net/books/24043/ (in Ukrainian).

4

Sustainable Development Goals as a Roadmap for Territorial Development

Liubov Zharova (Dr of Sci), Natalia Eremeeva (PhD)

4.1 Study Rationale

Sustainable development (SD) has become a popular catchphrase in contemporary development discourse. However, despite its pervasiveness and the massive popularity it has garnered over the years, the concept is still developing and transforming. The classical definition of sustainable development—this is the development that meets the needs of the present without compromising the ability of future generations to meet their own needs.[1.]

There are different points of view on appearing the concept of sustainable development. The main interpretations of sustainable development are eco-centric, on the one hand (which sets life, in general, and nature, in particular, at the center of the equation), and anthropocentric, on the other hand (the goal of which is mankind's wealth and survival). There is also a point, that we adopt the concept of SD because of a fear of using the term of "decrease" (as necessity to adopt a decrease of the used resources according to the fact that any productive activity induces an irreversible degradation of larger and larger quantities of energy and matter).[2]

Nevertheless the idea of "sustainable development" as a system of views of the ideologists in the Club of Rome, having emerged in the 1960–70s, generated agitated scientific and sociopolitical debates, involving the problems of the economy version, under which, based on the rational use of all resources, the use of new technologies and organizational management

[1] "Our Common Future" (1987) Report of the World Commission on Environment and Developmentю Online resource http://www.un-documents.net/our-common-future.pdf (accessed May 4, 2020)

[2] Sorin Avram Sustainable development - compromise or solution. What is the place of geography in this context? – Economics and Finance 15 (2014) 595 – 602

tools will secure the environment and the conditions for the development of the society in the future.

Traditionally we look at sustainable development as a balance between economic, social, and environmental parts of the development. Our first point of this research, that we should make the SD studies on the territory[3],[4] because we all live on territories that limited by the Earth and the nearest part of the Space around it (at least for this current moment), so far, any sustainable aspects should naturally include in the territorial focus, but we also need to take into consideration interconnections between territories and place of this territory in the geopolitical hierarchy. The second point is that we should not necessarily contrast eco- and socio-parts with economics.[5]

We should underline that SD issues are closely correlated with traditional economics ideas about the keystones of the economy—Scarcity, Choice, and Cost. Something is scarce if the choice of one alternative requires that another will be given up. The existence of alternative uses also forces us to make choices. The opportunity cost of any choice is the value of the best alternative forgone in making it that is why the latest research is moving to social from the mathematical spheres of economy. One more pillar of the economics is the idea that our resources are limited. That means that at any one time, we have only the defined set of resources of any kind (from natural to the virtual one), but our wants, our desires for the things that we can produce with those resources, are unlimited. We will always strive for more and better. Virtually everything is scarce.[6] This scarcity makes the

---

[3] Bhandari, Medani P. and Shvindina Hanna (2019) Reducing Inequalities Towards Sustainable Development Goals: Multilevel Approach, River Publishers, Denmark / the Netherlands- ISBN: Print: 978-87-7022-126-9 E-book: 978-87-7022-125-2

[4] Fluțăr Monica (2012). Sustainable development - a territorial development premise of Galați - Social and Behavioral Sciences 46 (2012) 1506 – 1509

[5] Bhandari, Medani P. (2019), Sustainable Development: Is This Paradigm The Remedy of All Challenges? Does Its Goals Capture The Essence of Real Development and Sustainability? With Reference to Discourses, Creativeness, Boundaries and Institutional Architecture, SocioEconomic Challenges, Volume 3, Issue 4, 97-128 ISSN (print) – 2520-6621, ISSN (online) – 2520-6214 https://doi.org/10.21272/sec.3(4).97-128.2019, http://armgpublishing.sumdu.edu.ua/wp-content/uploads/2020/01/9.pdf

Bhandari, Medani P. (2019), Inequalities with reference to Sustainable Development Goals in Bhandari, Medani P. and Shvindina Hanna (edits) Reducing Inequalities Towards Sustainable Development Goals: Multilevel Approach, River Publishers, Denmark / the Netherlands- ISBN: Print: 978-87-7022-126-9 E-book: 978-87-7022-125-2

[6] L. Zharova A. Chechel (2020). Historical Aspects of Sustainable Development and Economic Evolution Interconnection – Schid – N 2 (166), 2020 – P. 22-28 DOI: https://doi.org/10.21847/1728-9343.2020.2(166).201399

idea of sustainability actual and crucial for a long period, and the vitality of it based also on the world population forecast[7] which will also endeavor more.

That is why among all aspects of sustainable development, we concentrate on the territorial one. Thus, human existence is in a delimited territorial space. We have searched ways to harmonize human activities to preserve and conserve the ecological balance and operating realistically on the strategic direction of sustainable development based on analyses of the current situation in terms of its existing resources and opportunities for medium- and long-term use.

4.2 Methodological Background

We agreed that projecting the future is always a challenge and this exercise is not meant to be a prediction of a deterministic future. However, instead of trying to foretell what the future is likely to hold, models should be used to better understand how the world around us is unfolding given the path we seem to be on.[8] From that understanding, a policy should be made which considers the past, present, and possible future trajectories of sustainable human development. There are many ways to frame the uncertainty and test the reliability of projections including longitudinal statistical techniques, scenario analysis, historical forecasting, and qualitative analysis.

The methodological basis of sustainable development itself arise from technological and industrial development and growth of population that accompanied increase of prosperity and longevity which naturally led to rising of consuming and jeopardizing the natural resources' availability. Scientists and philosophers are of the understanding that human development should be in harmony with the world (like a noosphere[9]) more than

---

[7] Hans Rosling (2020). "How Did the World Population Change?" Gap Minder Online resource: https://www.gapminder.org/answers/how-did-the-world-population-change/ (accessed May 4, 2020)

[8] Jonathan D. Moyer, Steve Hedden, Frederick S. (2020). Are we on the right path to achieve the sustainable development goals? World Development 127 (2020) 104749

[9] Georgy S. Levit (2000). The Biosphere and the Noosphere Theories of V. I. Vernadsky and P. Teilhard de Chardin: A Methodological Essay – Online resource https://web.archive.org/web/20050517081543/http://www2.uni-jena.de/biologie/ehh/personal/glevit/Teilhard.pdf (accessed May 4, 2020)

pure economic development; they frightened with limits of growth[10] and tragedy of commons[11],[12] and warned about the consequences of uncontrolled growth.

By common belief, humanity's challenge in the 21st century is to meet the needs of all within the means of the planet. In other words, to ensure that no one falls short on life's essentials (from food and housing to healthcare and political voice), while ensuring that collectively we do not overshoot our pressure on Earth's life-supporting systems, on which we fundamentally depend—such as a stable climate, fertile soils, and a protective ozone layer. The Doughnut visualization of social and planetary boundaries is a playfully serious approach to framing that challenge, and it acts as a compass for human progress for this century (Figure 4.1)

Numerical endeavors to change the common understanding of the true price of economic development firstly touched the nongovernmental organizations then all together stimulated a rise in awareness of international organizations and governmental institutions and united them for mutual discussion and reviewing policies and politics, afterword it affected ordinary citizens and only then it caused to change the entrepreneurs' view of the situation.

Currently, most of the definitions and interpretations of sustainable development do not come from the comprehensive concepts of such development. Instead, they are influenced by the basic principles of specific

---

[10] Meadows, Donella H; Meadows, Dennis L; Randers, Jørgen; Behrens III, William W (1972). The Limits to Growth. A Report for the Club of Rome's Project on the Predicament of Mankind . New York: Universe Books.

[11] Hardin, G (1968). The Tragedy of the Commons. Science. 162 (3859): 1243–1248. Online resource http://science.sciencemag.org/content/162/3859/1243 (accessed May 4, 2020)

[12] Bhandari, Medani P, (2019). Live and let other live- the harmony with nature /living beings-in reference to sustainable development (SD)- is contemporary world's economic and social phenomena is favorable for the sustainability of the planet in reference to India, Nepal, Bangladesh, and Pakistan? Adv Agr Environ Sci. (2019);2(1): 37–57. DOI: 10.30881/aaeoa.00020 http://ologyjournals.com/aaeoa/aaeoa_00020.pdf

Bhandari Medani P. (2019). "BashudaivaKutumbakkam"- The entire world is our home and all living beings are our relatives. Why we need to worry about climate change, with reference to pollution problems in the major cities of India, Nepal, Bangladesh and Pakistan. Adv Agr Environ Sci. (2019);2(1): 8–35. DOI: 10.30881/aaeoa.00019 (second part) http://ologyjournals.com/aaeoa/aaeoa_00019.pdf

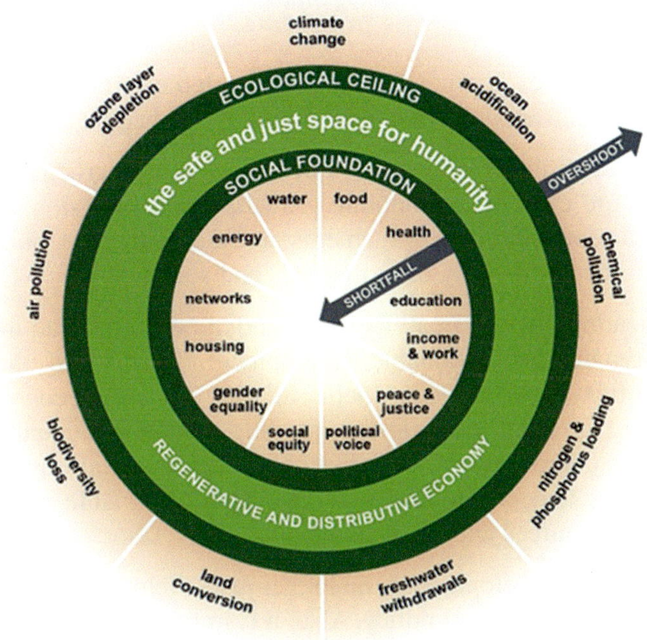

Figure 4.1 The Doughnut of social and planetary boundaries. [13]

groups or organizations of sustainable development.[14] Based on the study[15] of the evolution of sustainable thoughts and the formation of sustainable theory, evolution and development of SD theory could be divided into the embryonic period (before 1972), the molding period (1972–1987), and the developing period (since 1987). Nowadays, sustainable development has been defined as development that meets the needs of the present without

[13] Kate Raworth (2017). Doughnut economics: seven ways to think like a 21st-century economist. New York, United States: Random House. ISBN 978-184794138-1.

[14] Mebratu, D. (2005). Sustainability and sustainable development: Historical and conceptual review. Environ. Impact Assess. Rev. 1998, 18, 493–520. https://linkinghub.elsevier.com/retrieve/pii/S0195925598000195 ; Zhao, J.; Liang, X.; Zhang, X. System analysis on the definition of sustainable development. Acta Ecol. Sin. 1999, 19, 393–398; Kates, R.W.; Parris, T.M.; Leiserowitz, A.A. What is sustainable development? Goals, indicators, values, and practice. Environ. Sci. Policy Sustain. Dev. 2005, 47, 8–21.

[15] Mensah, J., & Ricart Casadevall, S. (2019). Sustainable development: Meaning, history, principles, pillars, and implications for human action: Literature review. Cogent Social Sciences, 5(1), 1653531

compromising the ability of future generations to meet their own needs. The main issues are[16]:

- Sustainable development calls for concerted efforts toward building an inclusive, sustainable, and resilient future for people and planet.
- For sustainable development to be achieved, it is crucial to harmonize three core elements: economic growth, social inclusion, and environmental protection. These elements are interconnected, and all are crucial for the well-being of individuals and societies.
- Eradicating poverty in all its forms and dimensions is an indispensable requirement for sustainable development. To this end, there must be promotion of sustainable, inclusive, and equitable economic growth, creating greater opportunities for all, reducing inequalities, raising basic standards of living, fostering equitable social development and inclusion, and promoting integrated and sustainable management of natural resources and ecosystems.

Thus, sustainable development is no longer just a field for scientific research; it is a framework for newest policy. Also, it is important to stress that sustainable development is not limited by ecological problems, resource devastation, and greening the economy; it is wider and more complex.[17] Summarizing, we could say that each step between milestones of sustainable development (1972: UN Conference on the Human Environment, Stockholm—1987: Brundtland Report—1992: UN Conference on Environment and Development, Rio de Janeiro—1997: Rio+5 Conference, New York—2002: UN World Summit on Sustainable Development, Johannesburg—2012: Conference on Sustainable Development "Rio+20"—2015 Agenda 2030…) were made not only because of political dash of countries (one asking for help and equality, others for dividing the presser of help equally) or because of awareness of the size and inevitability of the problems became clear for all but as a result of evolutionary changes in the economic development.

[16] The Sustainable Development Agenda (2020) Online recourse https://www.un.org/sustainabledevelopment/development-agenda/ (accessed May 4, 2020)

[17] Zharova L (2017) "Contemporary Managerial Theories Transformation in Framework of Sustainable Development" MIND Journal #4 (2017) Online resource https://mindjournal.wseh.pl/sites/default/files/article/03-18/contemporary_managerial_theories_transformation_in_framework_of_sustainable_development.pdf (accessed May 4, 2020)

Nowadays, the Roadmap for sustainable development is the Agenda 2030[18] aimed to stimulate action over the next 15 years in areas of critical importance for humanity and the planet and contains 17 goals and 169 tasks.

Consequently, the sustainable development program is a movement which captures the notion of development considering the three major pillars of development- economic, social and environment. It is mandated through the United Nations declaration, therefore, each UN member countries are obligated to include the sustainable development goals in their development planning.

The territory becomes a place where governmental ambitions met enterprises' needs. We cannot exclude any player from the field, but we should find a better way for communication and efficient implementation of SD ideas.

To illustrate this dilemma, we started with the upper level—country view on sustainable development. Till now almost every country has adopted 17 Goals and interpreted them to their needs, possibilities, understanding of the situation; and has designed the Agenda on how to move toward these goals. Naturally, every country demonstrates a different level of progress (Figure 4.2) and we can divide all of them due to traditional patterns (because it is actually very similar to the economical capabilities of the countries—Figure 4.3) on sustainably developed and countries that are only developing in a more sustainable way, we also can point the sustainably emerging countries—those who have some characteristics of great progress toward sustainable development but unevenly through all goals.

The issues are predictable—countries with better progress toward sustainable development not only have more economic resources, but they have a higher level of economic freedom i.e., reforming, raising awareness, and including stakeholders in the decision-making process are more effective and sometimes possible at all. Besides that, economically independent countries not waiting for solutions and resources, make their own way for SD and those who do not have enough resources, possibilities, and capabilities for doing this wait for help and resources, which means that they become hostages of wrong or inefficient decisions on all levels (country, world organizations, and international funds).

Hence, the country level is necessary but not sufficient part in the SD implementation. And for further analysis, we move to the enterprises level. There is need to discuss the business, which can become the stimulus (one of) in sustainable development.

---

[18] Transforming our world: the 2030 Agenda for Sustainable Development – Online resource – https://sustainabledevelopment.un.org/post2015/transformingourworld (accessed 04 May 2020)

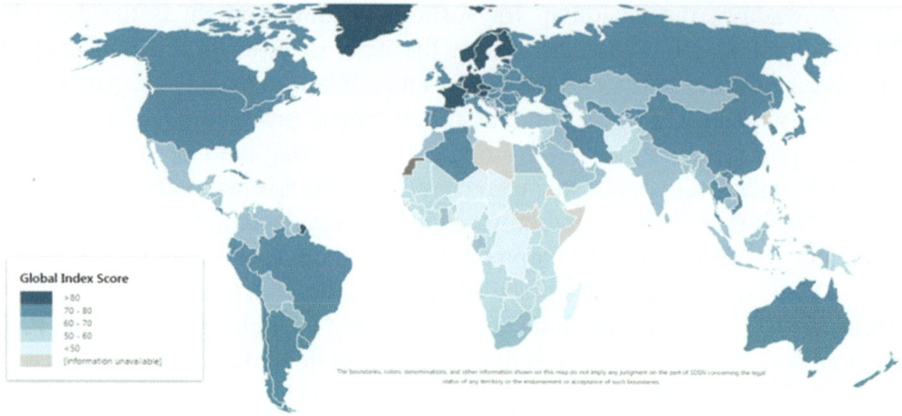

Figure 4.2 The ranking of countries by Sustainable Development Index (the higher score shows better progress and more sustainable development of the country).[19]

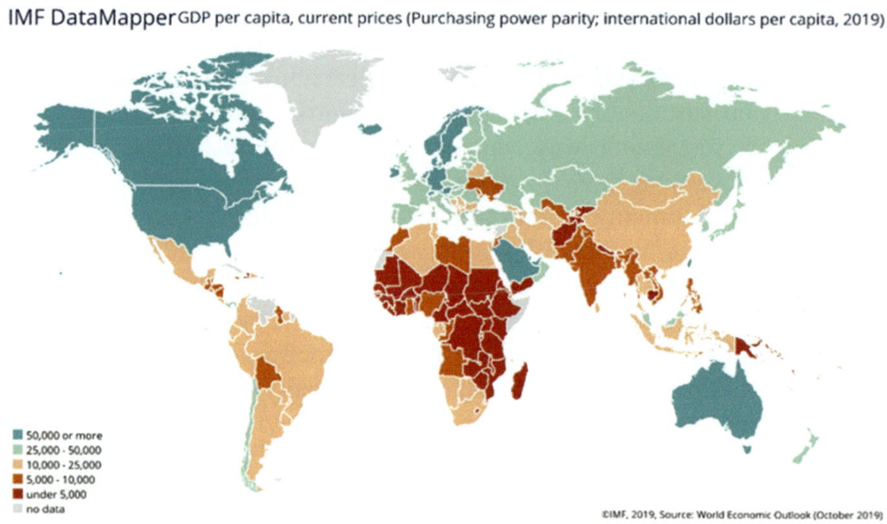

Figure 4.3 GDP per capita, current prices, 2019.[20]

The UN Agenda 2030 deems the private sector as pivotal in co-governing sustainability issues. Despite intense research on corporate sustainability, there is no explicit analysis on which policy-driven concepts companies

[19] Sustainable Development Report Dashboards (2019). Transformations to Achieve the Sustainable Development Goals. Online resource - https://dashboards.sdgindex.org/#/ (Accessed 24 May 2020)

[20] IMF Data Mapper - https://www.imf.org/external/datamapper/PPPPC@WEO/THA

should choose to forward their sustainability visions and practices. This is relevant because communication of corporate sustainability contributes to legitimizing or delegitimizing company actions, while simultaneously feeding back into public thinking and actions toward sustainability transformations. As per the content analysis[21] of DJSI (Dow Jonson Sustainability Index), global companies of three most common sustainability concepts mainstreamed at the global level: Circular economy (CE), Green economy (GE), and Bioeconomy (BE) shows that:

> CE be omnipresent and homogeneous across all companies and sectors;
>
> GE was the second most frequent concept, especially in forest and mining;
>
> BE was under-represented in all reports, with the exception of the forest sector. Interlinkages between concepts were few.

The CE–BE connection appeared to be the strongest, concerning efficiency and recycling of bio-based resources. The analysis of global sustainability concepts from the perspective of corporate disclosure enables a timely discussion on the role and limits of the business organizations as a participant to sustainability transformations globally but discloses the problem of interpretation of sustainability, ways of ensuring it, and business interests.

The Dow Jones Sustainability Index[22] family tracks the stock performance of the world's leading companies in terms of economic, environmental, and social criteria. All ratings evaluated in both 2012 and 2018 saw a significant favorable increase in expert perception of quality and changing priorities of goals for the business, in particular, in 2018 the most urgent issues become climate changes and quality of water and forest resources.[23] At the same time a more interesting picture presents the comparison of rating the same SD aspects by corporate and non-corporate representatives (Figure 4.4).

---

[21] D. D'Amatoa, J. Korhonena, A. Toppinena (2019). Circular, Green, and Bio Economy: How Do Companies in Land-Use Intensive Sectors Align with Sustainability Concepts? Ecological Economics 158 (2019) 116–133 https://doi.org/10.1016/j.ecolecon.2018.12.026; D'Amato, D., Droste, N., Allen, B., Kettunen, M., Lähtinen, K., Korhonen, J., Leskinen, P., Matthies, B.D., Toppinen, A., (2017). Green, circular, bio economy: a comparative analysis of three sustainability avenues. J. Clean. Prod. 168, 716–734. https://doi.org/10.1016/j.jclepro.2017.09.053

[22] Index Family Overview https://www.spglobal.com/esg/csa/indices/

[23] Rate the Raters (2019): Expert Survey Results February 2019 https://sustainability.com/wp-content/uploads/2019/02/SA-RateTheRaters-2019.pdf (Accessed 20 May 2020)

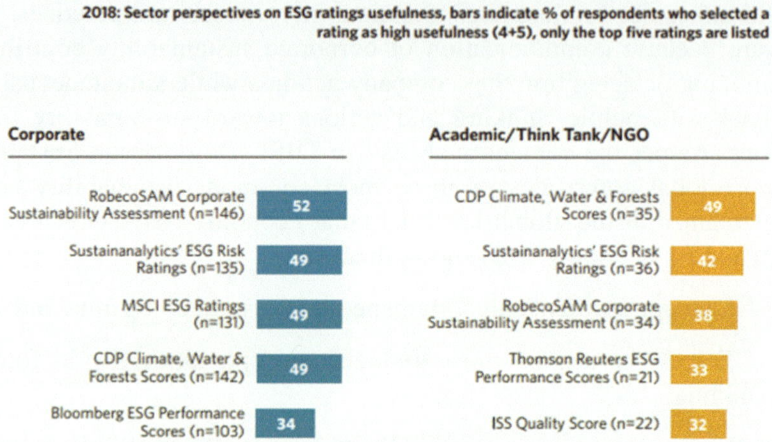

Figure 4.4 Corporate and non-business view on SD priorities.[24]

The idea of continued economic growth and business development and spreading in all spheres of life and geographical locations is increasingly questioned and critically analyzed based on its potential negative sustainability impact. In the face of pressing global social-ecological challenges,[25] private actors such as businesses are urged to engage in the voluntary governance of sustainability issues as key components of co-regulation, as envisioned, for instance, by the UN Agenda 2030 for Sustainable Development. Several scholars, however, have suggested that corporate sustainability would need more concrete visions and targets linked to fundamental sustainability issues in a holistic and inclusive manner.[26] Along with the critique, visions

[24] Rate the Raters (2019), Expert Survey Results February 2019 https://sustainability.com/wp-content/uploads/2019/02/SA-RateTheRaters-2019.pdf (Accessed 20 May 2020)

[25] MA - Millennium Ecosystem Assessment, (2005). Millennium ecosystem assessment synthesis report. In: Millennium Ecosystem Assessment. http://chapter.ser.org/europe/files/2012/08/Harris.pdf ;Rockstrom, J.E.A., Steffen, W., Noone, K., et al., (2009). A safe operating space for humanity. Nature 461, 472–475. https://doi.org/10.1038/461472a

[26] Addison, P.F.E., Bull, J.W., Milner-Gulland, E.J., (2018). Using conservation science to advance corporate biodiversity accountability. Conserv. Biol. 1–12. https://doi.org/10.1111/cobi.13190 ; Bjørn, A., Bey, N., Georg, S., Røpke, I., H., M.Z., (2016). Is Earth recognized as a finite system in corporate responsibility reporting? J. Clean. Prod. 106, 106–117. https://doi.org/10.1016/j.jclepro.2015.12.095 ; Lozano, R., Huisingh, D., 2011. Inter-linking issues and dimensions in sustainability reporting. J. Clean. Prod. 19, 99–107. https://doi.org/10.1016/j.jclepro.2010.01.004 ; Whiteman, G., Walker, B., Perego, P., (2013). Planetary boundaries: ecological foundations for corporate sustainability. J. Manag. Stud. 50, 307–336. https://doi.org/10.1111/j.1467-6486.2012.01073

and strategies for alternative systems need also be brought onto the agenda. The researchers[27] are continuously developing the multi-target back casting scenarios for illustrating future states in which four sustainability targets (climate, land use, participation, and resource security) are to be attained. The most common one that also based on "doughnut model," combination of Circular economy (CE), Green economy (GE), and Bioeconomy (BE), sharing economy in the realities of new wave of technological development:

Collaborative economy—citizen and governance prosumer networks

Circular economy in the welfare state—the government rules, supported by industry

Local self-sufficiency—household production in the locally governed communities

Automation for quality of life—robots produce while citizens enjoy leisure time

For stepping away from this methodological cul-de-sac, we stand on the fact that country vision and business interpretation of the sustainable development has met on the territorial level and the independence of local communities and the possibility of self-governance make sustainable development more effective and even realistic. From the other perspective, today's goals for local communities should be built on the SDG. Further research is the attempt to perform the practical ways of implementing our findings.

4.3 Strategic Vision and Territorial Development

The territorial strategies focus on promoting the sustainable development and improving the population's quality of life so that the inter- and intra-territorial

[27] Scenarios for sustainable futures beyond GDP growth 2050 - Åsa Svenfelt, Eva C. Alfredsson, Karin Bradley, Eléonore Fauré, Göran Finnveden, Paul Fuehrer, Ulrika Gunnarsson-Östlinga Karolina Isaksson, Mikael Malmaeus, Tove Malmqvist, Kristian Skånberg, Peter Stigson, Åsa Aretun, Katarina Buhr, Pernilla Hagbert, Erika Öhlund – Futures 111 (2019) 1–14; Sorin Avram Sustainable development - compromise or solution. What is the place of geography in this context? Procedia Economics and Finance 15 (2014) 595 – 602 http://creativecommons.org/licenses/by-nc-nd/3.0/ ; Ion Păun Otiman, Violeta Florian Ecosystems evolution as source of sustainable development - 1st International Conference 'Economic Scientific Research - Theoretical, Empirical and Practical Approaches', ESPERA 2013 - Procedia Economics and Finance 8 (2014) 327 – 335 ; Raworth, K. (2012). A safe and just space for humanity: Can we live within the doughnut?, Vol. 461https://doi.org/10.5822/978-1-61091-458-1 Oxford, UK

differences get attenuated; this desideratum can be materialized through the creation of conditions for each region to become a competitive territory on the long term and attractive for investments, with the development of the natural heritage, of the highly qualified human resources, creation of new labor employment opportunities. The territorial space must become an open and competitive territorial system with regional accessibility, mobility, and connectivity, through infrastructure modernization for ensuring a transport system that is able to provide fast and efficient connections to the regional and national markets, with the best use of the geostrategic positions of regions.

The territorial strategies by local communities enable the integrated development, in sustainable terms, according to the opportunities and constraints of rural communities and can ensure the necessary framework for the territorial, economic, and social cohesion. The general regional objective is to promote sustainable development and to improve the rural population's quality of life so that this becomes a competitive region in the long term, in which social cohesion should become an essential characteristic. The specific territorial objectives of the local strategies are determined by the economic and social specificity of each regional entity in part, by the evolutions and dynamics of inter-regional and intra-regional processes.[28]

For building up the strategy territorial development could use multi-criteria analysis models (MCAM)[29] which permit finding the way of balancing interests of different groups on multiple levels through Analytic Hierarchy Process (AHP),[30] which is focused on five fundamental steps:

---

[28] Ion Păun Otiman, Violeta Florian (2013). Ecosystems evolution as source of sustainable development - 1st International Conference 'Economic Scientific Research - Theoretical, Empirical and Practical Approaches', ESPERA 2013 - Procedia Economics and Finance 8 (2014) 327 – 335; Höjer, M., & Mattsson, L. G. (2000). Determinism and backcasting in future studies. Futures, 32(7), 613–634. https://doi.org/10.1016/S0016-3287(00)00012-4 ; Svenfelt, Å., Engström, R., & Svane, Ö. (2011). Decreasing energy use in buildings by 50% by 2050 — A backcasting study using stakeholder groups. Technological Forecasting and Social Change, 78(5), 785–796. https://doi.org/10.1016/j.techfore.2010.09.005

[29] Horizon/Europe (2020). Multi-Criteria Method Decision (MCMD): il metodo AHP [Giovanni Campeol, Sandra Carollo Nicola Masotto Infrastructural projects and territorial development in Veneto Dolomites: Evaluation of performances through AHP - 2nd International Symposium «NEW METROPOLITAN PERSPECTIVES» - Strategic planning,spatial planning, economic programs and decision support tools, through the implementation of Horizon/Europe2020. ISTH2020, Reggio Calabria (Italy), 18-20 May 2016 - Social and Behavioral Sciences 223 (2016) 468 – 474

[30] Maria Vișana (2019). Spatial and territorial development planning: digital challenge and reinvention using a multi-disciplinary approach to support collaborative work - Procedia Computer Science 162 (2019) 795–802 (7th International Conference on Information

- development of the hierarchy; in this first step the decision maker analyzes all the aspects of the problem and structures it in a hierarchy composed of several levels;
- elaboration of the pairwise comparison matrix; it involves an evaluation that identifies the weights to match with each criterion in the hierarchical problem, thanks to the use of an evaluation matrix whose single elements are obtained by pairwise comparisons of the problem criteria;
- determination of the relative local weights; once obtained the pairwise comparison matrix, in the subsequent step of the model the weights to be matched with each criterion are evaluated;
- analysis of the judgment consistency; in this step of the process it is necessary to verify whether the weights obtained in the previous step are consistent with the judgments expressed by the decision maker; and
- determination of the global weights: the principle of hierarchical composition; this is the final step and it involves calculating the global weights (or priorities) of the actions. In order to determine the importance of each element in relation to the Goal it is necessary to apply the principle of hierarchical composition.[31]

Considering these general scenario guidelines, the hierarchical model has been set up in order to apply it on the basis of the following characteristics or attributes, defined in relation to the scenario characteristics:

Society: effects on social demand and demography;
Landscape: catching and enjoying the landscape quality along the route;
Economy: promoting economic development in the territories stretching along the route;
Time: speed of overall travel times generated by the transport project;
Territory: possibility to communicate with the local transport network;
Costs: building costs, Project financing.

Interpretation of this approach to the purpose of our research is represented in Figure 4.5. Any level could not be ignored and at the same time we

Technology and Quantitative Management (ITQM 2019); Campeol G., Carollo S., Masotto N., (2015). Development Theories and Infrastructural Planning: the Belluno province. International Conference on Smart and Sustainable Planning for Cities and Regions 2015, Bolzano 19-20 November

[31] Saaty T.L., (1980). *The Analytic Hierarchy Process*. McGraw-Hill Book Co., New York

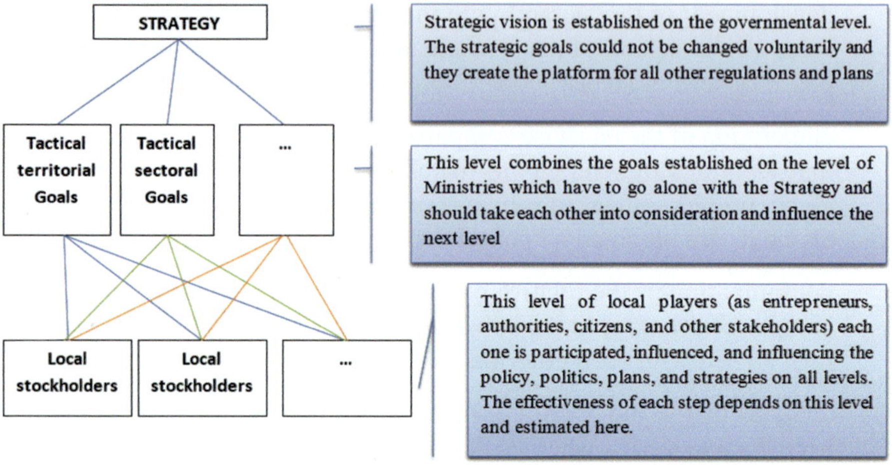

Figure 4.5 Diagram of dominance hierarchy in SDGs implementing process (developed by authors).

can observe that best governmental strategy will be ruined without support of all levels, and the basic players, including territories and local communities with their clear understanding goals and priorities are crucial.

The common problem is the segregation of these levels and the weak interconnection of players. Undoubtedly, the financial issues are always on time, but nevertheless the discussion of stimulation of local communities' development is much wider and it is also about understanding the weaknesses and opportunities of local communities for a long-term perspective and in the framework of sustainable development as a part of the main strategy, mentioned above.

In previous research,[32] we proposed a methodology for community analysis with the purpose of identifying the needs, wants, and gaps for strategy development.

Step 1. Scope—initial analysis based on the available data (geography, finances, including what taxes remain in the community, economic indicators, businesses, online reviews…). This step also includes the

[32] Zharova L. (2020), Sustainable Development in Framework of Smart City Initiatives // Economics and Law – N 3 (54) (2019). - http://economiclaw.kiev.ua/index.php/economiclaw/article/view/953 Accessed 01.05. DOI: https://doi.org/10.15407/econlaw.2019.03.107

estimation of consequences of climate change for the territory—because they vary in terms of geographical characteristics. Also, on this step, the local (regional) Networked Readiness Index could be developed based on the international methodology proposed by the World Economic Forum. The index referred to as Technology Readiness, measures the welfare for territories to exploit the opportunities offered by information and communications technology (ICT).

Step 2. Communication—the first stage of communication with stakeholders—what information is available and what (in their opinion) is missing, what are the expectations, what are the problems, how they see the opportunities and what they would like to have (money is not the answer, here we should try to understand their vision—agriculture, services, tourism…) and how we can mitigate the negative impact of climate changes or can we benefit in some areas from that changes.

Stages 1 and 2 analyze the community and identify the most promising areas of development and possible points of growth while assessing opportunities for implementation—infrastructure, finance, specialists (availability and who can be involved). It will also be clear what kind of information technologies can be applied now that need development.

If necessary, a few more meetings and consultations can be held to clarify and form a common opinion

Stage 3. Summarizing—the descriptive part is what we have now, and the promising part is what can be changed in the short term. It also includes a representing part—what can be offered to investors and a list of potential initiatives that the region may be interested in (manufacturing, international communities…).

Step 4. Action plan (to further develop the strategy)—it is a list of specific steps and their sequence to determine where to get money, who to contact, etc.

Territorial development policy envisages the economic growth of regions based on efficient use of own potential and resources, analysis of environmental aspects of regional development, and management. In the framework of the analysis of the experience of decentralization in Ukraine, data of the Organization for Economic Cooperation and Development to identify possible directions for improving the effectiveness of local sustainable development policy:

- Organization of communication with local entrepreneurs and collaboration with large companies operating in the territory based on the use of a cluster approach (or creating a cluster as a geographical concentration of similar, related or additional enterprises with active channels for business transactions, communications and dialogue sharing specialized infrastructure, labor markets and services, and share opportunities or threats);
- Stimulation of participation in various events (programs, fairs, meetings) for the development of territories of common interest, initiated by the association, by the state, the EU, etc. Such communications at high-level meetings are needed to agree on goals and goals for business development, identify key actors to support the goals set, and share experiences;
- Inspirational civic activism, education and outreach activities, fostering a culture of sustainability and society's demand for transparency, a clean environment, and social responsibility. Also, close cooperation in the form of dialogue with residents and entrepreneurs and managers of territorial associations will adequately assess the current skills needs for the development of small- and medium-sized businesses, as well as cooperation with large (regional or national) employers and the Universities and will give impetus to the development of institutions for professional education;
- Stimulating the development of start-ups, including training, the creation of incubators, etc. It is important to increase their English language skills for free access to best practices and opportunities for equal participation in international experience and discussion.

Thus, the discussion about how to reach the success on territorial level should be more specific we focus on Ukraine and Goal 7.

4.4 Choosing Priorities: Ukraine

One of the global problems of mankind is meeting the energy needs in the framework of the capacity of ecosystems, the attempt to solve which at the international level allows stating the formation of a set of legal methods of regulating environmentally-oriented energy, which is based on the concept of sustainable development.[33] The priority goals of regulation of this system of norms are rational resource use: ensuring environmentally friendly and prudent use of traditional energy resources and the use of renewable energy sources. For Ukraine, such a transformation of legal approaches to the regulation of relations in the energy sector is carried out in the context of European integration aspirations in accordance with the general principles of European energy space, defined in a concentrated form by the Energy Community Treaty and the Third EU Energy Package. It should be noted that the strategic priorities of the European energy area include not only sectoral goals (transformation of the economic system over time) but its territorial dimension—the spatial development of this system. The first group of priorities includes effective management of energy demand, energy efficiency, quality and diversification of supplies; ensuring the reliability of energy supplies to the EU; structural reform of national energy balances by reducing the share of natural gas and increasing the share of renewable energy sources in the overall structure of energy consumption; and improving energy security and environmental friendliness of energy production. The second group includes the development of local, regional raw material base, which in turn will help reduce the level of dependence on monopolists-suppliers of energy resources. The need to take into account the territorial (regional) aspect of sustainable development is emphasized by the EU's General Program of Environmental Action (paragraph 31), emphasizing that some challenges have a strong regional dimension.[34] Thus, the development of the energy sector is related to the conservation needs of lands, soils, and ecosystems (biodiversity) of certain areas.

The focus on the 7th goal is reasoned by the importance of energy safety and security issues for Ukraine. The energy security profile provides a brief snapshot of the energy security risks for Ukraine, including how it scores compared with the OECD average and a description of those factors that have had the greatest impact, both positively and negatively, on its energy security shows that country has one of the lowest levels of energy security (Figure 4.6.)

---

[33] Білоцький С.Д. Міжнародно - правові засади та інституційні механізми регулювання відносин в сфері екологічно орієнтованої енергетики: монографія. / С.Д. Білоцький. Одеса:Фенікс, 2015. – 646 с.

[34] Рішення № 1386/2013/EU Європейського Парламенту та Ради від 20 листопада 2013 р. про Загальну програму дій Союзу з охорони навколишнього середовища до 2020 року «Жити добре в рамках обмеженості ресурсів нашої планети» // Офіційний вісник Європейського Союзу. - 28.12.2013 р. - L 354/171

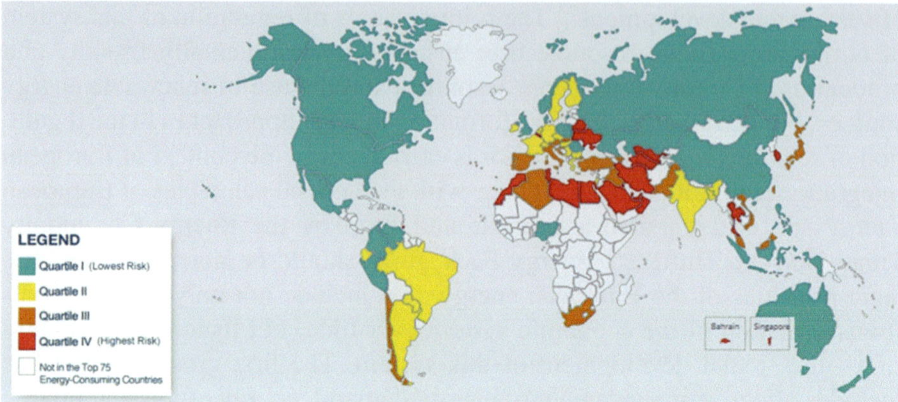

Figure 4.6 International Energy Security Risk Index.[35]

Among the threats to the state's energy safety could be highlighted the distortions of market mechanisms in the energy sector; insufficient level of diversification of energy sources and technologies; monopolization of energy industries by Russian capital; over-reliance on external markets; criminalization and corruption of the energy sector; its high "shadowing"; an inactive energy efficiency and energy policy; high deterioration of fixed assets of critical energy infrastructure facilities, and insufficient level of physical protection.[36]

There are no doubts that reforming the energy sector and its modernization for rising the energy safety and security should be done on the basis of the latest research, tendencies, and technologies in the framework of updated priorities not only in Ukraine but also in the EU with which the Association Agreement was signed.[37] It means that research should be held in the frame of sustainable development goals and to be more precise—Goal 7.

Coming back to strategic level as a framework for overall planning. The new State Regional Development Strategy for the period up to 2020 has been

[35] Global energy institute (2020) International Energy Security Risk Index. Available at https://www.globalenergyinstitute.org/international-energy-security-risk-index (accessed May 24, 2020)

[36] Рябцев Г.Л. (2019), Суходоля О.М. Нові загрози у сфері забезпечення енергетичної безпеки: моніторинг реалізації стратегії національної безпеки – Національний інститут стратегічних досліджень URL: https://niss.gov.ua/en/node/144 (accessed May 24, 2020)

[37] *Association Agreement* (2017) between the European Union and Ukraine. Bilateral institutions of the Association Agreement between Ukraine and the EU. Government portal. – URL: https://www.kmu.gov.ua/en/yevropejska-integraciya/ugoda-pro-asociacyu (Accessed 24 May2020)

adopted (approved by the Resolution of the Cabinet of Ministers of Ukraine on August 6, 2014, № 385). The integrated approach declared in the Strategy is based on the prioritized realization of own resource potential of the territory, and the spatial development grounded on the balanced development is through the formation of "growth points," strengthening the potential of rural areas, based on common management approaches to the formation and implementation of regional development policy, optimization of the territorial organization of government.

The main principles of state regional policy include sustainable development—like ensuring the inexhaustible, economical and efficient use of energy, material, natural and other resources to meet the needs of the current generation, taking into account the interests of future generations.

However, the available natural resources in the Strategy are not considered as an internal factor in the development of the region. The focus is on improving the natural environment only in order to achieve the recreational attractiveness of the region, and not on the efficient economic use of natural resources.

Thus, the Strategy actually lacks an active environmental component, as soon it is not harmonized with the Basic principles (strategy) of the state ecological policy of Ukraine for the period till 2020, its basic principles:

- the complexity of the environmental approach, which consists in solving the problems of the natural environment (atmosphere, water environment, soil, etc.);
- basin management principle;
- regulated distribution of environmental responsibility between the state, society, and business (corporations), documented in the relevant environmental policies;
- regulation (standardization) of environmental management and policy planning functions;
- systematic, comprehensive strategic environmental assessments;
- transfer of environmental protection powers to the regional and local levels.

Nowadays, in Ukraine (but the situation is not unique) we have a mess between all levels of decision-making and strategy implemented processes, that begins on the middle-level (Figure 4.5) of mutual agreement of steps and goals. Thus, The Ministry of Economic Development, Trade and Agriculture of Ukraine divided goals voluntarily for creating the SDG Report 2020 (Figure 4.7) that has raised questions and problems, e.g., Is the Goal 7 more environmental than economic for Ukraine? This can also distort the goal settings on all other levels and ruin the system.

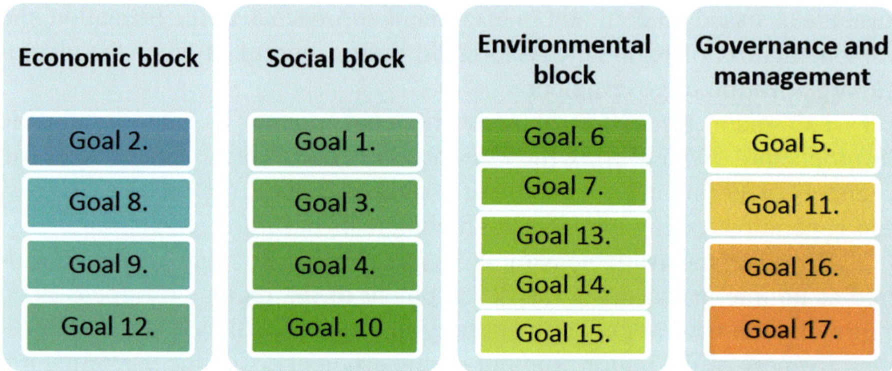

Figure 4.7. Dividing the goals by The Ministry.[38]

Goal 1. End poverty in all its forms everywhere.
Goal 2. End hunger, achieve food security and improved nutrition, and promote sustainable agriculture.
Goal 3. Ensure healthy lives and promote well-being for all at all ages.
Goal 4. Ensure inclusive and equitable quality education and promote lifelong learning opportunities for all.
Goal 5. Achieve gender equality and empower all women and girls.
Goal 6. Ensure availability and sustainable management of water and sanitation for all.
Goal 7. Ensure access to affordable, reliable, sustainable and modern energy for all.
Goal 8. Promote sustained, inclusive and sustainable economic growth, full and productive employment and decent work for all.
Goal 9. Build resilient infrastructure, promote inclusive and sustainable industrialization and foster innovation.
Goal 10. Reduce inequality within and among countries.
Goal 11. Make cities and human settlements inclusive, safe, resilient, and sustainable.
Goal 12. Ensure sustainable consumption and production patterns.
Goal 13. Take urgent action to combat climate change and its impacts.*

[38] https://www.me.gov.ua/Documents/Detail?lang=uk-UA&id=7e82b45d-3135-4788-99ce-f20ffbede635&title=PovidomlenniaProDraftDobrovilnogoNatsionalnogoOgliaduSchodoProgresuUkrainiUDosiagnenniTsileiStalogoRozvitku

Goal 14. Conserve and sustainably use the oceans, seas and marine resources for sustainable development.
Goal 15. Protect, restore, and promote sustainable use of terrestrial ecosystems, sustainably manage forests, combat desertification, and halt and reverse land degradation and halt biodiversity loss.
Goal 16. Promote peaceful and inclusive societies for sustainable development, provide access to justice for all and build effective, accountable and inclusive institutions at all levels.
Goal 17. Strengthen the means of implementation and revitalize the global partnership for sustainable development.

The environmental component of regional development in Ukraine includes such parts:

- the national system of environmental management as a system of the state, public, and corporate (business) environmental management has not been finally formed
- there is a state management system in the field of environmental protection, the authority, and functions of which are defined by the Law of Ukraine "On Environmental Protection"
- unregulated division of powers between local, regional, and central authorities

So, in the continuation of our ideas visualized in Figure 4.5 and results of our investigation we insist the strategic tasks of the first order—the distribution of powers among local, regional, and central authorities (Table 4.1).

The legal support can create this basis for designing the plans and establishing the goals, it is the pointer for local communities where they should seek for solutions. The legislation is also the kind of filtration grid that cut of inappropriate goals (e.g., those who conflicted with Strategy) and established measures in with communities can seek the efficient solutions.

That allows us formulate threats to the effective implementation of available tools:

- lack of a unified methodology for the formation of targeted programs, including the violation of the relationship between the purpose, activities, performers, and resources of the program;
- inactive use of regional development agreements (out of 27 regions, such agreements have been concluded only with 6 oblasts: Donetsk (completed in 2011), Lviv, Volyn, Vinnytsia, Ivano-Frankivsk, and Kherson)
- • lack of rules for ranking programs by priority (competition of programs marked as a priority, which makes it impossible to finance them)

Table 4.1 The distribution of powers authorities (developed by authors)

Level of AUTHORITIES	FUNCTIONS
Central	• prevention of ecosystem pollution; • harmonization of interaction between society and nature, reproduction of ecological balance; • ecological rehabilitation and ecological safety.
Regional	• regulation of the use of natural resources of local significance; • determination of standards for environmental pollution (installation of GDV, GDS, and waste disposal); • introduction of the economic mechanism of nature management; monitoring and accounting of objects of nature use and environmental pollution; • conducting state ecological expertise; • state control over compliance with environmental legislation; • development of programs for the implementation of environmental measures, definition, and implementation of investment policy; informing the population and interested institutions and organizations on environmental issues.
Local	• conducting local and object monitoring; • monitoring compliance with environmental legislation; • organization of development of local ecological programs and projects.

- lack of financial resources, the imperfection of the organizational and economic mechanism of program activities;
- unformed system of control over the implementation of target programs.

Nevertheless, the opportunities for territorial development now are to use available tools for environmental policy implementation: regional environmental programs; regional development agreements; comprehensive regional environmental programs. Opportunities or the main ways to direct regional (territorial) policy to solve environmental problems that may be implemented in the nearest future:

- making direct investments in environmental infrastructure, in particular, water treatment facilities;
- conducting Strategic Environmental Assessments (SEAs) for relevant programs; environmental impact assessments in the preparation of all major projects;

- stimulating the participation of representatives of environmental groups and agencies in the development of programs and control over them.

4.5 Conclusions

Lastly, we would like to emphasize the fact that the country vision and business interpretation of the sustainable development have met on the territorial level and the independence of local communities and the possibility of self-governance make sustainable development more effective and even realistic. From the other perspective, today's goals for local communities should be built on the SDGs.

The territorial strategies by local communities enable the integrated development, in sustainable terms, according to the opportunities and constraints of rural communities and can ensure the necessary framework for the territorial, economic, and social cohesion. The general regional objective is to promote sustainable development and to improve the rural population's quality of life so that this becomes a competitive region in the long term, in which social cohesion should become an essential characteristic. The specific territorial objectives of the local strategies are determined by the economic and social specificity of each regional entity in part, by the evolutions and dynamics of inter-regional and intra-regional processes.

Also, the researchers identify the main functions of territorial intermediation (and their specific elements) that local authorities could have to encourage the deployment and success of these projects.[39] Local authorities play the role of intermediation by (i) ensuring spatial and cognitive proximities between actors, (ii) mobilizing territorial resources and favoring local anchorage (iii) installing trust among the local stakeholders, and (iv) having a role of instigator by participating in the supply (inputs) and the purchase of the (outputs).

Summarizing for Ukraine, let's say that in an effective territorial development should be built in the context of SDGs formulated on the governmental level and can stimulate or accelerate such process like:

- raise the standards of living in the community and increase the level of employment;

[39] Bourdin, Sebastien (2020). Fabien Nadou The role of a local authority as a stakeholder encouraging the development of biogas: A study on territorial intermediation - Journal of Environmental Management - Volume 258, https://doi.org/10.1016/j.jenvman.2019.110009

- create a platform for dialogue and involvement of the population in discussing problems (forming a public inquiry) and finding solutions, especially in issues very urgent but unpopular (or at least not understandable in short run) like combating climate changes;
- attracting investment in territory and creating new businesses through increasing transparency of the processes and cutting the time for starting projects and business; and
- creation of a positive image of the territory in the country on the international market, which can give impetus for economic development.

Territorial development could be interpreted as the basis of local self-government and the designing of an effective system of financial capacity at different levels, since the national economy is formed also at the basic territorial level with using the information technology. Therefore, forming an effective model for the territorial development required considering the specifics of interactions of the local economy and urgent national (regional) economic processes. Undoubtedly, one of the urgent processes is the choice of specialization, the other—design a territorial brand based on the use of territorial advantages and resources. Resources are an integral value and a unique component that combines natural, human, infrastructural, and more.

The current situation in Ukraine with the legal support of secondary resource use is characterized by a non-systemic and segmented approach to the regulation. The legislation of Ukraine in this area is an array of patchworked rules and regulations. The determination of adequate economic and legal methods of secondary resource usage in energy should build on the comprehensive approach to the legal regulation of these relations based on the method of legal economic order (the optimal combination of market self-regulation of economic relations and state regulation of macroeconomic processes—Article 5 Civil Code of Ukraine[40]). Within this approach, the public interest is provided by economic and legal methods of macroeconomic influence. At the national level—by forming state development programs, improving the regulatory framework of the energy sector of the economy, improving permitting procedures and licensing in this area.

Legislative support on the modern phase of Ukrainian development should include:

[40] 7. Господарський кодекс України. Відомості Верховної Ради України. 2003. № 18, № 19-20, № 21–22. Ст.144

- modernization of the legislation on the planning of development of regions and cities—transition to ecological, social and economic planning of development of regions and cities;
- adoption of the Laws “On Strategic Environmental Assessment,” “On Environmental Assessment of Projects”;
- adaptation of the Law “On Public-Private Partnership” for the purposes of regional environmental policy; and
- creation of legislative bases for assessment of natural resource potential of regions, development of normative-legal base concerning ecological and economic macro-regions; classification of regions according to the levels of technogenic and ecological risks, and creation of appropriate data banks.

So far, territorial development is a complex issue that could become the platform for overcoming after lockdown times and stimulate the economic growth of the country, because the seeking for a local solution could be simpler than reforming and stimulating the growth on the governmental level. The government should create the framework based on the SDGs and territories would find their innovative way to prosperity (and Goals 7 can give then new understanding of resources and opportunities for development).

5

Financialization as a Determinant of Income Inequality: The Case of Eastern European Countries

Tetyana Vasilyeva[1], Serhiy Lyeonov[1], Olena Pakhnenko[1], Pavlo Rubanov[1]

[1] Sumy State University, Ukraine
Corresponding author: pakhnenko@gmail.com

ABSTRACT

The financial sector plays an important role in society and the economy. First, the financial sector provides a secure money savings. Second, it simplifies payments and other transactions. In addition, risk insurance is provided by the appropriate financial intermediaries. Finally, the financial sector plays an intermediary role in redistributing savings and channeling them into investment and real sector financing. The current stage of global economic development is characterized by financialization—an increase in the importance of financial motives and in the number of financial players, financial intermediaries, financial institutions in operations in the domestic and foreign markets; an increase in the share of the financial industry in the world economy, financial controllers in corporate management, financial assets among other assets, securities in circulation and shares among other financial assets; the stock market tends to be a market for corporate control in defining the corporate strategy and fluctuations in the financial markets to determine the business cycles.

On the one hand, the development of the financial system is a positive phenomenon, which in the case of its reliability and stability guarantees the provision of all the necessary financial services, including meeting the needs of vulnerable group of population, such as people living in poverty. However, on the other hand, studies show the opposite effect of financial sector development, which is characterized by increasing inequality in the distribution of economic income. This has several manifestations. For example, financial products reward the wealthy and cost the poor, increasing the inequality gap; financial sector has moved into extracting rents from the real sector; capital markets system rewards short-term profit maximization instead of sustainable business practices, etc.

The chapter analyzes the development of income inequality, driven by the growing role of the financial sector in the economy. The paper uses Eastern European countries as a case study, and provides a comparative analysis of the level of financialization in developed European countries and the Eastern European countries with lower levels of economic development, as well as a comparison of the impact of financialization on income inequality in the respective country groups.

In addition, given the current development of the financial sector, which is under the influence of digitalization and is characterized by an increase in the technologization of financial services, the authors also examine the relationship between these phenomena, in particular between the level of digitization of the economy and the level of its financialization, and the level of technologization of financial services and income inequality.

5.1 Introduction

The structural proportions of the development of modern economies are characterized by an increase in the share of services in the volume of gross product, and in the services sector, in turn, financial services play an important role. In general, the patterns of increasing the role of the financial sector in the economy, growth of GDP generated in this segment, expanding the sphere of influence and diversification of the institution of financial intermediation, etc. are summarized by the concept of financialization. The main characteristics of financialization are:

- an increase in the importance of financial motives and in the number of financial players, financial intermediaries, and financial institutions in operation in the domestic and foreign markets;
- an increase in the share of the financial industry in the world economy, financial controllers in corporate management, financial assets among other assets, securities in circulation, and shares among other financial assets;
- the stock market tends to be a market for corporate control in defining the corporate strategy and fluctuations in the financial markets tend to determine the business cycles.

The functioning of the financial sector and the activities of financial intermediaries have a multifaceted impact on economic processes. The role of the financial sector is significant in ensuring the efficiency of economic processes, servicing current financial transactions, redistribution of capital between individual entities of the economy and its industries, and financial risk management (Fernandes, 2018; Vasiljeva & Stadnyk, 2019; Bhandari 2020; Bhandari and Shvindina 2019). First, the financial sector provides a secure money savings. Second, it simplifies payments and other transactions. In addition, financial intermediaries provide risk insurance and risk hedging instruments. Finally, the financial sector plays an intermediary role in redistributing savings and channeling them into investment and real sector financing.

Financialization and its consequences are the subject of research in many scientific papers. At the same time, the conclusions of these studies regarding the positive and negative effects of financialization are extremely contradictory. The target vector of development of modern economies is sustainable development, which includes, among other components, economic growth and welfare. In this context the development of the financial system is a positive phenomenon, which in the case of its reliability and stability guarantees the provision of all the necessary financial services, including meeting the needs of vulnerable group of population, such as people living in poverty (Morsher et al., 2017; Singh, 2018).

At the same time, a number of scientific studies substantiate the thesis that there is not always a straightforward relationship between financialization, economic growth, and the welfare of the population. In other words, the growth of the financial sector leads to increased inequality in the distribution of income between economic agents. This has several manifestations. For example, financial products reward the wealthy and cost the poor, increasing the inequality gap. Another manifestation researcher sees that is the financial sector has moved into extracting rents from the real sector. Besides, capital markets system rewards short-term profit maximization instead of sustainable business practices (Hardoon & Shigiya, 2017; Storm, 2018).

In a financial economy, much of the profit-making activity has nothing to do with the real sector. Increasing the capitalization of the stock market through speculative transactions carries additional risks of financial bubbles and financial crises (Balas et al., 2019; Khan et al., 2017). The risk factor for financial crises is income inequality and the credit boom, as confirmed by the 2008 crisis. At the same time, the opposite thesis is true that the poorest segments of the population feel the most negative consequences of global financial crises, which further deepens the problem of financial inequality (Dave, 2017).

Contradictory conclusions about the relationship between the level of financialization and income inequality are complex economic systems; the end result is determined by the combined influence of many different factors. Most often, the additional factors considered in the analysis of the correlation between financialization and income inequality are the type of financial system and the level of economic development.

For example, Maldonado (2017) hypothesizes that the direction of the impact of financialization on income inequality depends on the type of financial system of the country—bank-oriented or market-oriented financial system. Bank-oriented models are characterized by the dominance of the banking sector and lending as a method of financing. Market-oriented models are based on the dominant role of the stock market and are characterized by the predominance of placement of corporate securities over bond loans. The classic examples of market-oriented models of financial systems are the United States and the United Kingdom, while the financial systems of France and Germany are bank-oriented. For the most part, European countries tend to bank-oriented models. However, in its pure form, bank-oriented and market-oriented models of financial system do not exist, which makes it difficult to distinguish their impact as a factor in changing income inequality due to financialization. Thus, one part of the research finds that the growth of the banking component in the financial system has a negative impact on the equality of income distribution (Maldonado, 2017). Proponents of another concept take the view that the stock market creates more opportunities for higher income for capital owners by investing it in securities, which has a much stronger impact on increasing income inequality compared to the bank-oriented model (Brei *et al.*, 2018).

Another factor that should be considered when analyzing the interdependence of financialization and income inequality is the general level of economic development (Jafarzadeh & He, 2019). For example, in developing countries, financial sector growth and credit expansion can have positive effects on income equalization if loans are used for high-yield investments, including education and employment. For developed economies, this issue is less relevant given the availability of these benefits due to higher levels of household income and government support (Courn'ede et al., 2015).

Among the studies describing the correlation between income inequality and credit expansion depending on the level of financial development, it is worth mentioning the paper by Greenwood and Jovanovic (1990), in which the authors propose to use a U-shaped relationship. Thus, financial development and credit expansion first reduce inequality, and when a certain limit is reached, they change the sign of correlation to the opposite.

An empirical study of the impact of financial development on income distribution, conducted on the example of African countries for the period 1990–2004, showed a reduction in income inequality with the development of the financial sector (Batuo *et al.*, 2010). The study also showed the influence of the factor of the level of education on the uniformity of income distribution. Demirgüç-Kunt and Levine (2009) also confirmed the positive impact of financialization on income equalization in developing countries.

At the same time, studies of modern financial development in developed countries often show the growth of financial inequality and the creation of greater economic benefits from financialization for higher wage classes. Thus, the Rajan study (2010) notes that growing income inequality in the United States and the lack of wage growth in the run-up to the global financial crisis of 2008 stimulated the use of credit resources by low- and middle-income households to maintain current consumption. This, in turn, provided a redistribution of income in favor of the richer class, increasing inequality.

At the present stage in the study of income inequality and financialization it is necessary to take into account the existence of another defining trend—the growth of informatization of society and the transition to Industry 4.0 (Kuzmenko & Roienko, 2017).

Given the regional, economic, political, social, and other differences in the development of countries, the object of study of this paper are countries belonging to the common region of Eastern Europe and characterized by a relatively close level of economic development (Kostel et al., 2017; Kouassi, 2018; Pilia, 2017). The objectives of this study are to analyze the level of financialization in this group of countries and a comparative analysis with other European countries; construction and analysis of regression dependences of income inequality on the level of financialization; and the study of the correlations between the level of digitalization of the financial sector, and income inequality and the level of financialization. The relevant three sections of the paper are devoted to these components.

5.2 Methods

The paper uses quantitative indicators of financial sector development, digitization and income sharing in Europe, and particularly in Eastern Europe countries. We assess the overall level of financialization and study its dynamics for the period 1996–2017. We use open-source data from the World Bank and Eurostat. The methods of comparative, correlation, and regression analysis are used to assess the degree of dependence of income inequality on financial sector development indicators in different groups of countries, as well as correlation analysis methods to identify the relationship between the level of digitization of financial sector and indicators of financialization and income inequality.

5.2.1 Comparative Analysis of Financialization in Eastern Europe and Other European Countries

At the first stage of the study we conduct the comparative analysis of the level of financialization in developed European countries and the Eastern European countries with lower levels of economic development.

The overall level of financial sector penetration into the economy (financial depth) is usually analyzed by indicators that reflect the relative size of market and bank components, namely the weights of stock market capitalization and bank assets relative to gross domestic product (GDP).

In the paper the total volume of the financial sector or the overall level of financialization is proposed to be estimated as the sum of the ratios of banks' assets to GDP and the capitalization of the stock market to GDP. According to research, in OECD countries both the size of the total financial sector and stock market capitalization have increased significantly since the 1970s. The authors analyze this indicator for European countries in 2017 (Figure 5.1).

From the obtained results it is possible to draw some conclusions. First, the average level of the overall size of financial sector to GDP in 2017 in European countries is 146%. At the same time, all analyzed countries of Eastern

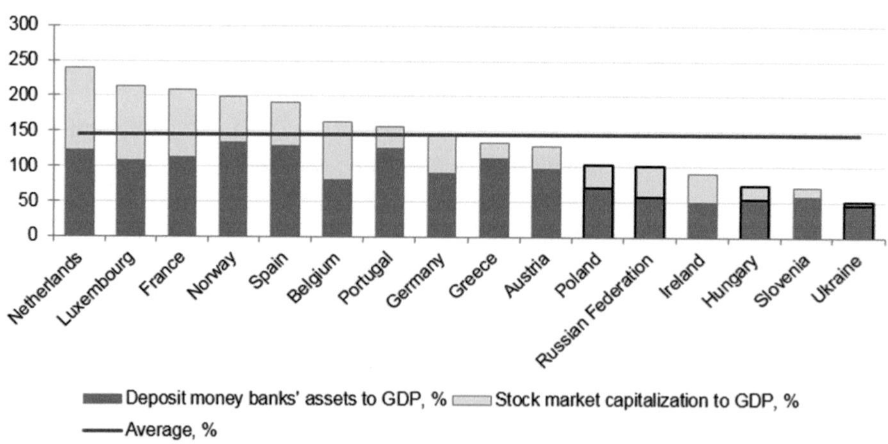

Figure 5.1 Overall size of financial sector in European countries in 2017.
Source: Own adjustment based on World Bank data

Europe (Poland, Russia, Hungary, Ukraine) have an actual value of the size of the financial sector that is significantly lower than the European average. For example, Poland–103%, Russia–101%, Hungary—74%, Ukraine—50%. Moreover, Eastern European countries occupy the last positions by the level of financialization along with other European countries characterized by a relatively lower level of economic development (Ireland—92%, Slovenia—71%). The highest level of financialization is typical for the countries of Northern Europe—the Netherlands (239%) and Luxembourg (214%).

The second conclusion from the analysis of the overall size of the financial sector in European countries is the different ratio between the contribution of the banking sector and the stock market to the overall financialization rate. In some countries (Netherlands, Luxembourg, France, and Belgium) the contribution is almost equal or with the predominance of the stock market. The rest of the countries are characterized by a larger share of the banking sector. At the same time, in such countries as Portugal, Greece, Hungary, Slovenia, and Ukraine, the banking sector is dominant in the structure of the financial system. The obtained results reflect the model of the financial system formed in these countries—bank-oriented or market-oriented.

At the same time, a formed model of the financial system cannot be considered permanent. Under the influence of financial globalization there is a convergence of models of financial systems with the formation of mixed features. On the other hand, each country has its own characteristics of the financial system, regardless of the dominant model. Due to the influence of economic and non-economic factors, certain regulatory decisions, the ratio of the banking sector and the stock market may change, as well as the model of the financial system may be transformed as a whole.

This thesis is confirmed by data on the dynamics of changes in the level of financialization in European countries during 1996–2017 (Figure 5.2).

The cyclical changes in the average level of financialization in Eastern Europe and other European countries for the period 1996–2017 are similar. At the same time, there are significant differences in trends, especially in recent years. In particular, during 1996–2000 and 2003–2007 there was a dynamic to increase the level of financialization in all European countries, including the Eastern European region. However, the growth rate of financialization was higher in developed European countries. As a result of the financial crisis in 2008 and subsequent years, there was a reduction in the level of financialization of European countries. However, the rate of reduction of financialization in developed European countries was sharper compared to Eastern Europe.

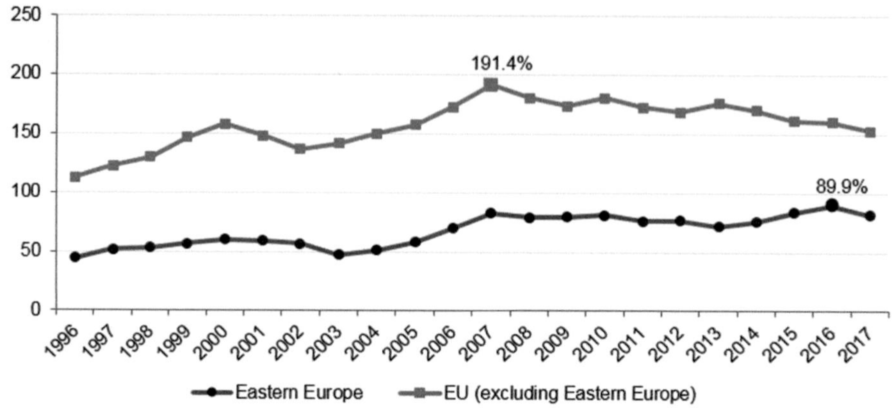

Figure 5.2 Dynamics of the average level of financialization in European countries during 1996–2017, %.
Source: Own adjustment based on World Bank data

In 2007, European countries (excluding Eastern Europe) reached the highest level of financialization for the analyzed period—the average rate for these countries was 191,4%. After that, there was a general tendency to a gradual decrease (except for 2010 and 2013, when there was a slight increase in the average value of the indicator). But in general, by 2017, the average level of financialization in the EU fell to 153,2%. At the same time, in Eastern European countries the dynamics of change of the indicator had insignificant deviations and changed mainly in the range of 72–82% after 2007. In 2016, the maximum value of the analyzed indicator for Eastern European countries was reached, namely 89,9%.

Thus, after the crisis of 2008, changes in the level of financialization in Eastern European and other European countries showed significant differences in the direction and the strength of their fluctuations. However, there is still a significant gap between the average value of the level of financialization in Eastern Europe (82,1% in 2017) and other European countries (153,2% in 2017). That is, in general, the level of financialization in developed European countries is almost twice as high as the average for Eastern European countries.

5.2.2 Financialization and Income Inequality Regression Analysis

In the next stage of the study, we conducted a regression analysis of the relationship between the level of financialization and the level of welfare of the population expressed in terms of the ratio of bank assets to GDP and the Gini coefficient. The Gini coefficient is a widely used and recognized inequality measure (Nagy & Kiss, 2018). As for the financialization, two indicators are most often used to analyze the level of financialization—bank assets to GDP and stock market capitalization to GDP. The choice depends on the type of financial system—bank-oriented or market-oriented. But the type of financial system is not clearly and permanently fixed for the country. Market transformations that occur under the influence of internal and external factors can lead to a shift in the direction of increasing the role of banks or the stock market. It is also influenced by legislative initiatives and crisis processes. Given the predominance of the banking sector in most of the countries studied, we chose the ratio of banking assets to GDP as an indicator of financialization. The results of the regression analysis of the impact of the level of financialization on the inequality of income distribution are presented in Table 5.1.

Table 5.1 Regression analysis results between financialization and income inequality

Countries with a statistically significant relationship between parameters				**Countries with unconfirmed significance of the relationship between parameters**	
Country	**Regression coefficient**	**Determination coefficient**	**Bank assets to GDP in 2017**	**Country**	**Bank assets to GDP in 2017**
Sweden	0,07146	0,69857	135,3	Denmark	172,3
Spain	0,03165	0,3899	129,9	United Kingdom	131,7
Italy	0,02239	0,31323	118,1		
France	0,10703	0,55487	113,0	Portugal	125,9
Greece	0,03188	0,51358	112,3	Austria	97,1
Luxembourg	0,05906	0,34024	109,1	Germany	91,1
Belgium	0,06713	0,28322	80,9	Slovak Republic	68,5
Norway	–0,04268	0,29371	134,4		
Netherlands	–0,16820	0,60404	123,3	Bulgaria	60,9
Finland	–0,02154	0,57446	100,3	Slovenia	59,0
Poland	–0,14961	0,6774	70,1	Ireland	50,8
Estonia	–0,05821	0,29752	67,4	Latvia	46,9
Czech Republic	–0,03932	0,30659	61,3	Belarus	45,9
Russian Federation	–0,05850	0,76327	58,1	Lithuania	44,6
Hungary	–0,11810	0,36758	54,3		
Ukraine	–0,08923	0,7622	46,4	Moldova	24,5
Romania	–0,09414	0,4136	36,7		

The study of European countries for the period 1996–2017 by the methods of correlation and regression analysis allowed dividing the countries into three groups. The first group includes countries that are characterized by a close relationship between the studied indicators and the change of these indicators is directly related, i.e., there is a positive correlation between the indicators, the change of one indicator leads to a unidirectional change of another. Accordingly, the second group of countries is characterized by a close statistically significant relationship between indicators, but the correlation is negative, i.e., the relationship is inverse: an increase in one indicator leads to a decrease in another and vice versa. And the third group of countries are countries for which no statistically significant relationship was found between the studied indicators.

The first group of countries with a direct correlation (increase in the level of financialization leads to an increase in income inequality) includes highly developed European countries, in which the ratio of banking assets to GDP is the highest (more than 80%), i.e., further growth of the banking sector in these countries leads to an increase in income gaps.

The second group of countries is characterized by an inverse relationship between the indicators, i.e., the growth of financialization leads to a reduction in income inequality. Such patterns have been found in countries where, firstly, the share of the financial sector is low, and secondly, the level of economic and financial development of these countries is lower. The growth of assets of the banking sector in such countries is evidence of economic development, improving the financial system, growing demand for financial services, attracting more people to service, increasing the deposit and lending activities of banks. This has a positive effect on the well-being of the population. According to the results presented in Table 5.1, all Eastern European countries with a statistically significant relationship between the level of financialization and the Gini coefficient belong to the second group. They showed the presence of an inverse correlation between the studied indicators. The same dynamics is typical for other European countries that have a relatively lower level of economic development (for example, Estonia).

The third group of countries for which there is no statistically significant relationship between the studied indicators are countries with a developed banking sector, but the ratio of banking assets to GDP is relatively low, so that the level of financialization is not excessive and does not have a significant impact on income equality.

It is worth noting that in some countries that have a high ratio of bank assets to GDP, no statistically significant relationship was found, but these countries have a different type of financial system.

Three countries were identified as countries that have an inverse type of relationship between the level of financialization and the Gini coefficient, but also a high level of financialization and a high level of economic development. These include Norway, the Netherlands, and Finland. Thus, the Netherlands has a high ratio of bank assets to GDP, as well as a significant (not lower) rate of stock market capitalization to GDP (for example, 116% in 2017). Significant capitalization of the stock market and other criteria allow to refer the Netherlands to the countries with a market-oriented model of the financial system. It does not follow the patterns of direct impact of financialization on the strengthening of financial inequality. Finland also has the characteristics of a market-oriented model of the financial system, although after 2008 its share of banking assets increased, while the capitalization of the stock market decreased significantly.

Thus, we can conclude that in countries with a developed and dominant banking sector and high financial performance (the share of bank assets in GDP is over 100%), the hypothesis of a negative impact of financialization on income distribution is confirmed. Countries with lower levels of economic development, including most Eastern European countries, have the opposite trend. In these countries, the growing share of the banking sector is evidence of economic development and contributes to rising middle-class incomes and reducing inequality.

5.2.3 Correlation Analysis of Digitization, Financialization, and Gini Index

After the financial crisis of 2008–2009, another trend in the development of the global financial system was the growth of digitalization and the beginning of technologization of the financial sector. The financial crisis led to a decline in the assets of the banking sector and the crisis in the stock markets. Further growth of the banking sector slowed down. The need to restore the stability of the banking system in the post-crisis period, lack of resources for investment in technological innovation, loss of confidence of households and businesses in traditional financial intermediaries and their reorientation to alternative market participants, strengthening government regulation of the financial sector and other economic factors have weakened the market position of traditional financial intermediaries (Goncharenko, 2018; Logan & Esmanov, 2017).

At the same time, a significant trend was the growth of technological financial services and the development of the FinTech segment as an alternative to traditional financial intermediaries. From a technological point of view, the basis of this transformation of the financial sector was the beginning of the release of smartphones, which created the opportunity to provide financial services through mobile applications. Technology companies and startups, not financial institutions, began to develop such applications. In other words, the development and complexity of information technology has led to the emergence of both new financial services and innovative financial technologies, and new players in the financial services market—FinTech companies (Rizwan & Semenog, 2017).

Therefore, today it is not expedient to analyze financialization as the only component and feature of the financial sector development and its impact on the economy and economic entities (Louis, 2017). Given the current development of the financial sector, which is under the influence of digitalization and is characterized by an increase in the technologization of financial services, the authors also examine the relationship between these phenomena, in particular between the level of digitization of the economy and the level of its financialization, and the level of technologization of financial services and income inequality.

The results of the correlation analysis in different countries showed very different results in terms of correlation between the level of financialization and digitalization of banking services, as well as digitalization and the Gini coefficient. In addition, no common pattern was found for the group of developed European countries and countries with a lower level of economic development, which includes the countries of Eastern Europe. The results of correlation analysis for countries for which a statistically significant correlation was established for one or both pairs of the studied parameters are presented in Table 5.2.

Characterizing the current stage of development of the financial sector of the national economy, we can identify several following major trends. First, the processes of digitalization of the financial sector continue, which at the current stage are manifested in the increasing share of online banking and mobile banking in the structure of banking services. This is due to an increase in the number of Internet users and the share of consumers who use their smartphone to receive financial services; increasing the quality of the Internet and mobile coverage, ensuring their availability; expanding the user's capabilities in online banking, increasing the range of financial transactions that can be conducted online (Agnihotri & Gupta, 2019).

Second, the FinTech services market continues to grow, with new applications and new areas of FinTech innovation. While in the initial stages of development, payments and transfers were of the greatest interest to FinTech companies, now FinTech companies are gradually entering the markets and

Table 5.2 Correlation analysis results

Countries	Internet banking in 2017, %	Growth rate of Internet banking from 2006 to 2017	Correlation with Gini index, coefficient*	Correlation with financialization rate (bank assets to GDP), coefficient*
Eastern Europe:				
Bulgaria	5	400,0	0,6721	Not significant
Czech Republic	57	470,0	–0,7133	0,6936
Hungary	38	375,0	0,8128	–0,6851
Poland	40	344,4	–0,8527	0,9515
Slovak Republic	51	292,3	Not significant	0,9339
Other Europe:				
Belgium	67	139,3	–0,7179	–0,7433
Denmark	90	57,9	0,9453	Not significant
Finland	87	38,1	–0,8830	0,9761
France	62	244,4	Not significant	0,9079
Germany	56	75,0	Not significant	–0,9697
Ireland	58	176,2	Not significant	–0,8294
Italy	31	244,4	0,8756	0,7621
Latvia	61	177,3	–0,6758	Not significant
Luxembourg	76	85,4	0,6497	Not significant
Netherlands	89	50,8	–0,7095	0,6042
Norway	92	37,3	Not significant	0,8962
Portugal	31	210,0	-0,7695	Not significant
Spain	46	206,7	0,6901	Not significant
Sweden	86	50,9	0,7791	0,9178

other types of financial services: investment, savings, insurance, personal finance management and more. Users of FinTech services are also increasing the list of FinTech services they use. Third, banks' awareness of the growing role of FinTech companies in the financial services market results in changes in banking business models, approaches to customer engagement, and consideration of collaboration with FinTech companies (Christensen et al., 2015; Poliakh & Alikariyev, 2017).

At the same time, the uneven penetration of digitalization and the uneven development of the FinTech segment in different countries at the present stage do not allow to identify significant correlations between the processes of digitalization of the economy and indicators of financialization and income inequality. However, we cannot exclude the growing interdependence between these indicators in the future while achieving higher indicators of development of the FinTech sector and strengthening the integration of banks and FinTech companies.

5.3 Conclusion

For a long time, financialization has been a defining trend in the development of the financial system of developed countries. Since the second half of the 20th century, the volume of bank assets and stock market capitalization have shown rapid growth and their ratio to GDP has reached 191,4% in developed European countries. After the crisis of 2008, the growth rate of financialization slowed down and the further increase in the share of banking assets and stock market capitalization in GDP remained only in countries with relatively lower levels of economic and financial development. Countries that have previously reached a high level of financialization, on the other hand, have shown some reduction in financialization.

At the same time, since 2008 the economic situation has contributed to the strengthening of digitalization and the beginning of technologization of the financial system and the emergence of FinTech companies in its structure, which have taken over some of the functions of traditional financial intermediaries. This is one of the reasons for the slowdown in financialization in developed countries.

An analysis of the relationship between the level of financialization and financial inequality in European countries confirmed the results of studies on the growth of inequality with increasing levels of financialization, but only in those countries that have achieved a high level of financialization and have bank-oriented model of the financial system. Eastern European countries, in contrast to this hypothesis, have shown an inverse relationship between financialization and income inequality, i.e., the growth of the financial sector leads to equalization of incomes of economic entities, due to the low level of development of financial systems in these countries. The increase in the level of financialization in these countries is accompanied by the development of the financial sector and the expansion of the middle-income class.

As for the current trend of technologization of the financial sector and increasing digitalization in the banking system, the study showed the lack of a clear pattern of the relationship between digitalization and financialization, and between digitalization and income distribution. Thus, we can say that digitalization is the only general trend common to all European countries, regardless of their level of economic development. The difference between groups of countries is the current level of digitalization, which in Eastern European countries is significantly lower. At the same time, no significant correlation was found between the parameters of digitalization, financialization, and income distribution (financial inequality).

Acknowledgement

The survey was supported by the Ministry of Education and Science of Ukraine and performed the results of the projects 0118U003569 and 0120U102001.

References

1. Agnihotri, A., Gupta, S. (2019). Relationship of Corporate Governance and Efficiency of Selected Public and Private Sector Banks in India. *Business Ethics and Leadership, 3(1)*, 109-117. DOI: http://doi.org/10.21272/bel.3(1). 109–117. 2019
2. Balas, A.N., Kaya, H.D. (2019). The Global Economic Crisis and Retailers' Security Concerns: The Trends. *SocioEconomic Challenges, 3(2),* 5-14. DOI: http://doi.org/10.21272/sec.3(2). 5–14. 2019.
3. Batuo, M., Guidi, F., Mlambo, K. (2010). Financial Development and Income Inequality: Evidence from African Countries. *MPRA Paper 25658,* University Library of Munich, Germany. URL: https://mpra.ub.uni-muenchen.de/25658/1/MPRA_paper_25658.pdf
4. Bhandari, Medani P. (2020) Second Edition- Green Web-II: Standards and Perspectives from the IUCN, Policy Development in Environment Conservation Domain with reference to India, Pakistan, Nepal, and Bangladesh, River Publishers, Denmark / the Netherlands. ISBN: 9788770221924 e-ISBN: 9788770221917

5. Bhandari, Medani P. (2020), Getting the Climate Science Facts Right: The Role of the IPCC (Forthcoming), River Publishers, Denmark / the Netherlands- ISBN: 9788770221863 e-ISBN: 9788770221856
6. Bhandari, Medani P. and Shvindina Hanna (2019) Reducing Inequalities Towards Sustainable Development Goals: Multilevel Approach, River Publishers, Denmark / the Netherlands- ISBN: Print: 978-87-7022-126-9 E-book: 978-87-7022-125-2
7. Brei, M., Ferri, G., Gambacorta, L. (2018). Financial structure and income inequality. *BIS Working Papers, 756.* Bank for International Settlements, 2018. URL: https://www.bis.org/publ/work756.pdf
8. Cherowbrier, J. (2015). European banking sector. A statistical dossier on the current banking sector and meeting the demands of a digital future. Statista, 2018. URL: https://senat.me/wp-content/uploads/study_id59063_european-banking-sector.pdf
9. Christensen, C. M., McDonald, R., Raynor, M. (2015). What Is Disruptive Innovation? *Harvard Business Review, 12,* 44–53. URL: https://www.hbs.edu/faculty/Pages/item.aspx?num=50233
10. Cournède, B., Denk, O., Hoeller, P. (2015). Finance and Inclusive Growth. *OECD Economic Policy Papers, 14.* OECD Publishing, Paris. URL: https://www.oecd-ilibrary.org/economics/finance-and-inclusive-growth_5js06pbhf28s-en
11. Dave, H. (2017). An inquiry on social issues – Part 1. *Business Ethics and Leadership, 1(2),* 78-87. Doi: 10.21272/bel.1(2). 78–88. 2017
12. Dave, H. (2017). An Inquiry on Social Issues – Part 2. *Business Ethics and Leadership, 1(3),* 45-63. DOI: 10.21272/bel.1(3). 45–63. 2017
13. Demirgüç-Kunt, A., Levine, R. (2009). Finance and inequality: Theory and evidence. *Annual Review of Financial Economics, 1,* 287-318. DOI: http://doi.org/10.1146/annurev.financial.050808.114334.
14. Eurostat. *Digital economy and society: Database.* URL: http://ec.europa.eu/eurostat/web/digital-economy-and-society/data/database
15. Fernandes, A. J. L. (2018). Transnational banks' influence on the development of the economy and the financial sector of developing countries (on the example of Poland, Brazil, Turkey). *Marketing and Management of Innovations, 3,* 253-259. DOI: http://doi.org/10.21272/mmi.2018. 3–22
16. Goncharenko, T. P. (2018). Current transformations of banking management: strategic analysis of external factors of direct influence activities. *Marketing and Management of Innovations, 2,* 365-377. DOI: http://doi.org/10.21272/mmi.2018. 2–28

17. Greenwood, J., Jovanovic, B. (1990). Financial development, growth, and the distribution of income. *Journal of Political Economy, 98(5),* 1076–1107.
18. Hardoon, D., Shigiya, K. (2017). Financing Inequality. *Future of Horizons Conference, April 2017.* Financial Conduct Authority. URL: https://www.fca.org.uk/publication/research/financing-inequality.pdf
19. Jafarzadeh, E., He, Sh. (2019). The Impact of Income Inequality on the Economic Growth of Iran: An Empirical Analysis. *Business Ethics and Leadership, 3*(2), 53-62. DOI: http://doi.org/10.21272/bel.3(2). 53–62. 2019.
20. Khan, K., Qingyang, W., Khurshid, A. (2017). Causal Relationship between Monetary Policy and the Stock Market: A Bootstrap Rolling Window Approach. *Financial Markets, Institutions and Risks, 1(4),* 5-15. DOI: 10.21272/fmir.1(4). 5–15. 2017
21. Kostel, M., Leus, D., Cebotarenco, A., Mokrushina, A. (2017). The Sustainable Development Goals for Eastern Partnership Countries: Impact of Institutions. *SocioEconomic Challenges, 1(3),* 79-90. DOI: 10.21272sec.1(3). 79–90. 2017
22. Kouassi, K. B. (2018). Public Spending and Economic Growth in Developing Countries: a Synthesis. *Financial Markets, Institutions and Risks, 2(2),* 22–30. DOI: 10.21272/fmir.2(2). 22–30. 2018
23. Kuzmenko, O., Roienko V. (2017). Nowcasting income inequality in the context of the Fourth Industrial Revolution. *SocioEconomic Challenges, 1(1)*, 5-12. DOI: 10.21272/sec.2017. 1-01
24. Logan, W., Esmanov, O. (2017). Public financial services transparency. *Business Ethics and Leadership, 1(2),* 62-67. DOI: 10.21272/bel.1(2). 62–67. 2017
25. Louis, R. (2017). A new economic order for global prosperity. *SocioEconomic Challenges, 1(2),* 52-58. DOI: 10.21272/sec.1(2). 52–59. 2017
26. Maldonado, J. E. (2017). *Financial Systems and Income Inequality. Bruges European Economic Research Papers, 36/2017. URL: http://aei.pitt.edu/id/eprint/85962*
27. Morsher, Ch., Horsch, A., Stephan, J. (2017). Credit Information Sharing and Its Link to Financial Inclusion and Financial Intermediation. *Financial Markets, Institutions and Risks, 1(3),* 22-33. DOI: 10.21272/fmir.1(3). 22–33.2017
28. Nagy, Z. B., Kiss, L. B. (2018). The Examination of Appearance of Income Inequality in Scientific Databases with Content Analysis.

Business Ethics and Leadership, 2(4), 35–45. DOI: http://doi.org/10.21272/bel.2(4). 35-45.2018.
29. Pilia, G. (2017). Estonia and Lithuania in transition: A compared analysis of the change and its costs and benefits. *Business Ethics and Leadership, 1(2),* 12-19. DOI: 10.21272/bel.1(2). 12–19.2017
30. Poliakh, S., Alikariyev, N. (2017). Evaluation Quality of Consumer Protection by Financial Markets Services. *Financial Markets, Institutions and Risks, 1(3),* 75-81. DOI: 10.21272/fmir.1(3). 75–81.2017
31. Rajan, R. (2010). Fault Lines: How Hidden Fractures Still Threaten the World Economy. Princeton University Press.
32. Rizwan, Ch A., Semenog, A. (2017). Non-bank financial institutions activity in the context of economic growth: cross-country comparisons. *Financial Markets, Institutions and Risks, 1(2),* 39-49. DOI: 10.21272/fmir.1(2). 39–49.2017
33. Singh, S.N. (2018). Regional Disparity and Sustainable Development in NorthEastern States of India: A Policy Perspective. *SocioEconomic Challenges, 2(2),* 41-48. DOI: 10.21272/sec.2(2). 41–48.2018
34. Storm, S. (2018). Financialization and Economic Development: A Debate on the Social Efficiency of Modern Finance. *Development and Change, 49(2),* 302-329. DOI: https://doi.org/10.1111/dech.12385
35. The World Bank. *DataBank.* URL: https://databank.worldbank.org/data/databases
36. Vasiljeva, T., Stadnyk, A. (2019). Innovative Approaches to Evaluation of Concentration of the Banking System as a Basis of Improving the State Crisis Management. *Marketing and Management of Innovations,* 1, 302–313. DOI: http://doi.org/10.21272/mmi.2019. 1–25

6

Poverty and Inequality in Access to Housing in Mexico City During Neoliberalism

Carlos Sanchez[1]

PhD Candidate in Economics from the Institute of Economic Research of the National University of Mexico

ABSTRACT

This work provides a scalar explanation of the current conditions of poverty and inequality of the population of Mexico City caused by the lack of access to housing, understood as a chain of events unleashed by neoliberalism as the development path undertaken by Mexico from the eighties of the twentieth century: deindustrialization, depopulation, and poverty. While carrying out an exercise to measure and map poverty and inequality due to homelessness of the inhabitants of the 16 counties and functional areas of Mexico City as an exercise to find public policy responses to one of the most important challenges in the great metropolis.

6.1 Introduction

In the rise of globalization, neoliberalism eroded Mexico City as a central scale in order to join the supranational scope without the direct mediation of the State. This distorted the urban in the City, because of a set of chained events: deindustrialization, depopulation, and poverty. Thus, the rise of the

[1] PhD Candidate in Economics from the Institute of Economic Research of the National University of Mexico and Second place in the Postgraduate and Researchers category of the National Research Award 2014 "Boosting development of state finances", awarded by National University, Interacciones financial group and PricewaterhouseCoopers Mexico. Email: carlirius@gmail.com

market as a representative of the social brought with it a pattern of uneven development, affecting the production of housing and distorting its public policy. The withdrawal of the State and its institutional framework places the inhabitants of Mexico City in different degrees of poverty and social inequality induced by their lack of housing according to the area of the city they inhabit.

Therefore, this chapter includes two objectives developed in its two sections. The first is to understand the chained events that shaped differential access to housing in Mexico City and its 16 counties grouped into three functional areas—Central City, which includes four counties and is the most economically dynamic nucleus of the City; six adjacent counties that embrace the central nucleus; and six more counties or Southern Counties, which represent the largest area of the City and house its rural area. The second objective is to measure and map the differentiated impacts that the set of events has had on the conditions of poverty and inequality of the population caused by lack of housing in the 16 counties of the City and its functional areas.

The results presented here warn about the need to return to the role of the State in the stewardship of public housing policy based on a new institutional framework that strengthens its actions in order to balance territorial inequalities in access to the housing of the inhabitants of Mexico City.

6.2 Deindustrialization, Depopulation, and Poverty: The Way Neoliberalism Shaped Housing Policy in Mexico City

Until before the decade of the eighties of the twentieth century, Mexico sustained a development model based on industrialization by import substitution with an important State intervention. Once the crisis of the eighties began, the economic model restructured and with it the social political orientation of the State, which had an impact on the population, as well as on the housing policy and poverty conditions of the country's inhabitants. Arguing that it was necessary to deregulate the sectors of the economy to boost capital accumulation and thereby recover from the crisis, the market answer to the social needs was neoliberalism: trade liberalization, economic theory of comparative advantages, and the minimum State.

The undertaking of a neoliberal development path in Mexico involved the elimination of subsidies to the population and the privatization of pensions, turning this fund into a system of individual capitalizations managed by private entities. The application of these measures, far from improving the situation of families, exacerbated poverty and conditions of economic inequality in several sectors. The abandonment of the self-sufficiency strategy for development radically transformed Mexico's relationship with other economies worldwide, moving from self-reliance to dependence, accompanied by the signing of the North American Free Trade Agreement (NAFTA) (Boltvinik, 2012). Since 1982, the deregulation strategy imposed the idea that the living standards of the population would improve through the liberation of the markets, the suppression of State intervention, and the economic takeoff derived from the commercial opening. Neoliberalism placed Mexico as the country with the largest number of commercial treaties signed in the world, which encouraged the progressive dismantling of the limited national industry and sectoral control instruments.

Federal public investment fell from 10.4% of GDP in 1982 to 4.9% in 1988, and 2.5% in 2002. Public spending on general economic sector development (concept that includes agricultural, manufacturing, and energy sector development) was reduced from 11.9% of GDP in 1982 to 8.7% in 1988, and 3.7 % in 2002 (Calva, 2004). Although this liberalization had an impact on the increase in exports from 14% of GDP to 25.3% in 2002, this improvement did not reflect in economic growth, since during this period, GDP only increased 0.56 times, at an average annual rate of 2.3%. In contrast, during the industrialization by import substitution strategy (1935–1982), GDP increased 15.9 times, at an average annual rate of 6.1% (Calva, 2004).

In this context, deindustrialization accompanied by economic liberalization played a key role in the increase or tertiary activities of the economy of Mexico City and a new population dynamic at the federal and local levels, which modified the objectives of housing and social development policies. Thus, from 1980 onwards, the Mexican population continued to grow with a dynamic of deceleration and demographic control, from which it went from 66 million inhabitants in 1980, to 81 million in the 1990s, and 91 million in 2000, to 103 million in 2010 (INEGI, 2017).

At the local level, until the mid-1970s Mexico City was still a privileged scale in the new industrial establishments, it was until the 1980s that the deindustrialization process reflected in the economic variables. This slow but progressive deindustrialization of the City is the result of a trend of relocation of the industry in the metropolitan area, accompanied by a readjustment in the spatial distribution of the population.

Although this dynamic did not end with the total expulsion of the industry, it influenced to diminish it considerably, since some City counties maintained an important participation in certain activities of the industrial sector, oriented largely to supply the local and national market. With this process, there was a growth in the activities of the tertiary sector such as commerce and services. From 1980 to 1988, although at a national level there was an increase in activities of the industrial sector in 3.8%, this did not reflect in Mexico City, where that growth was barely just 0.03% per year (Williams, 1997).

Deindustrialization had an impact on all counties in Mexico City: Cuauhtemoc as a central county deindustrialized in almost all subsectors, although the infrastructure conditions of the county led to the persistence of industrial activity like printing presses, publishing houses, and clothing manufacturing. Miguel Hidalgo as a central county too and Azcapotzalco as an adjacent county have a strong industrial trajectory that also lost industry participation in several branches apart from printing and publishing, accompanied by a moderate growth of the wood industry and products derived from it. Gustavo A. Madero and Iztacalco as an adjacent county had a loss in the concentration of jobs in the industrial sector, however, they recorded an upward participation of the garment and plastic products, cellulose, paper and derivatives manufacturing industries. The adjacent county of Iztapalapa maintained the metal products industry, the chemical industry, the paper industry, printers and publishers, in accordance with the trend in other counties. Between 1988 and 1994 there was an increase in jobs in the branch of the food industry (mainly bakery products), and the textile industry. While in the central county of Benito Juarez, deindustrialization progressed, maintaining the publishing industry. Finally, the southern counties that maintained an industrialization process were Tlahuac and Xochimilco, which in 1994 saw an increase in industrial employment, Tlahuac specializing in the industrial sector and Xochimilco in the chemical and pharmaceutical industry. This dynamic of location of the activities had consequences in the population derived from the type of activity concentrated in each county, recording important variations per decade.

This deceleration scenario expresses population dynamic in Mexico City as follows: the population growth rate remained on average below 3%, with 2.5% from 1980 to 1990; 3.04% from 1990 to 1995; 1.27% from 1995 to 2000; and 1.49% from 2000 to 2010. This dynamic is explained by the growth of tertiary activities of the economy in these decades, which tends to disperse the demographic component, in addition to the industry moving to the outskirts of the City, encouraging a dispersion of the demographic concentration as shown in Table 6.1.

Because of a re-densification policy in Mexico City between 2000 and 2006, there was a recovery of the population growth rate in the central counties from –17.9% in 1990 to 2.6% in 2010. In the adjacent counties, there is a process of contraction of the population growth rate, due to the combination of a recovery process of the growth rate in some counties such as Gustavo

Table 6.1 Population evolution by counties in Mexico City 1990–2010

County or functional area	1990	1995	2000	2005	2010
Cuauhtemoc	595,960	540,382	516,201	521,348	531,831
Miguel Hidalgo	406,868	364,398	352,640	353,534	372,889
Venustiano Carranza	519,628	485,623	462,806	447,459	430,978
Benito Juarez	407,811	369,956	360,478	355,017	385,439
Central counties	1,930,267	1,760,359	1,692,125	1,677,358	1,721,137
Azcapotzalco	474,688	455,131	441,008	425,298	414,711
Gustavo A. Madero	1,278,888	1,256,913	1,235,083	1,193,161	1,185,772
Alvaro Obregon	640,749	676,930	686,783	706,567	727,034
Iztapalapa	1,490,499	1,696,609	1,768,785	1,820,888	1,815,786
Coyoacan	640,066	653,489	640,423	628,063	620,416
Iztacalco	448,322	418,982	411,288	395,025	384,326
Adjacent counties	4,973,212	5,158,054	5,183,370	5,169,002	5,148,045
Xochimilco	265,131	327,290	358,223	404,458	415,007
Tlalpan	484,544	552,395	581,397	607,545	650,567
Tlahuac	209,013	257,572	308,786	344,106	360,265
Magdalena Contreras	195,041	211,898	221,979	228,927	239,086
Cuajimalpa	114,385	136,873	151,222	173,625	186,391
Milpa Alta	66,524	84,100	99,437	115,895	130,582
Southern counties	1,334,638	1,570,128	1,721,044	1,874,556	1,981,898
Total, Mexico City	8,238,117	8,488,541	8,596,539	8,720,916	8,851,080

Source: Census information INEGI.

A. Madero, Iztacalco, and Azcapotzalco; while there is a decrease in other counties such as Iztapalapa, Alvaro Obregon, and Coyoacan [see Charts 6.1 and 6.2]. In the case of the southern counties of Mexico City, a reduction in the population growth rate from 40.2% in 1990 to 9.6% in 2000, and 5.7% in 2010 is noticeable [see Chart 6.3].

Thus, each level of industrialization corresponds to a degree of centralization and demographic concentration that not necessarily establishes in the industrialized zone. In this way, the following population analysis periods for Mexico City are established: the first goes from 1950 to 1970 and describes the growth of the industrial activity in the "central city," which stimulated population growth in some counties with greater industrial participation, accompanied by a depopulation in counties with little influence in the industry. The second describes a trend of relative decrease, which goes from 1970 to 2000, where there is a demographic rearrangement in a certain parallel with the deindustrialization and boom of the activities of the tertiary sector of the City's economy. During this period, some counties noted negative growth rates. Finally, a third period from 2000 to 2010, where it is possible to distinguish a dynamic of re-densification, accompanied by a concentration

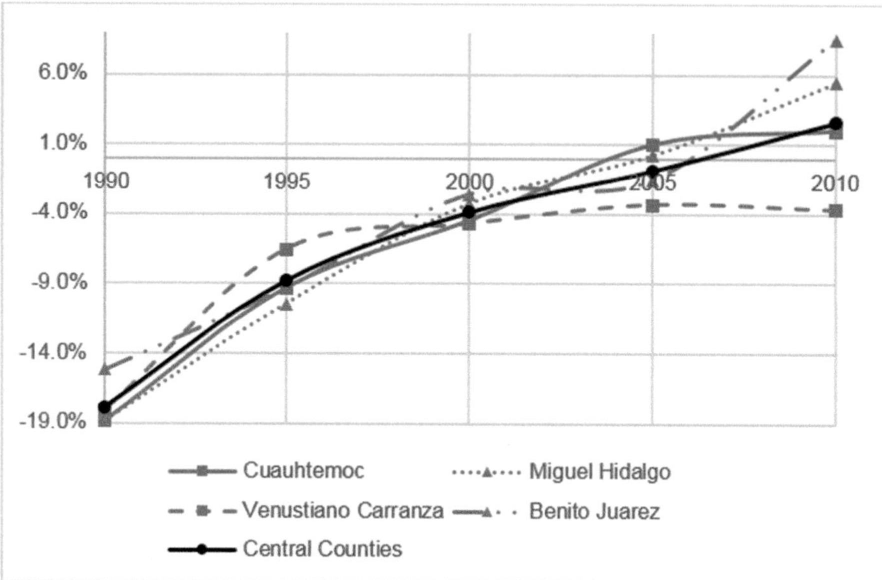

Chart 6.1 Evolution of the population growth rate in Mexico City (1990–2010), Central Counties
Source: Own elaboration with data from Table 6.1

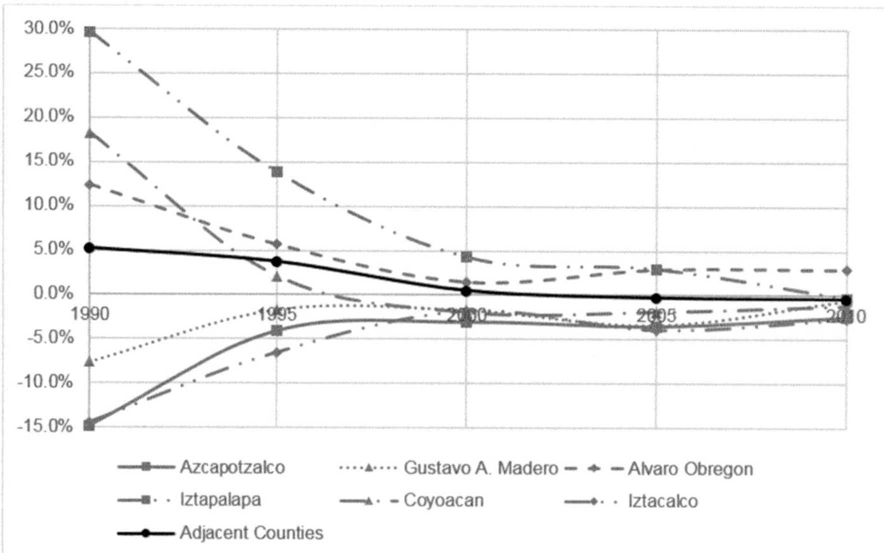

Chart 6.2 Evolution of the population growth rate in Mexico City (1990–2010), Adjacent Counties
Source: Own elaboration with data from Table 6.1

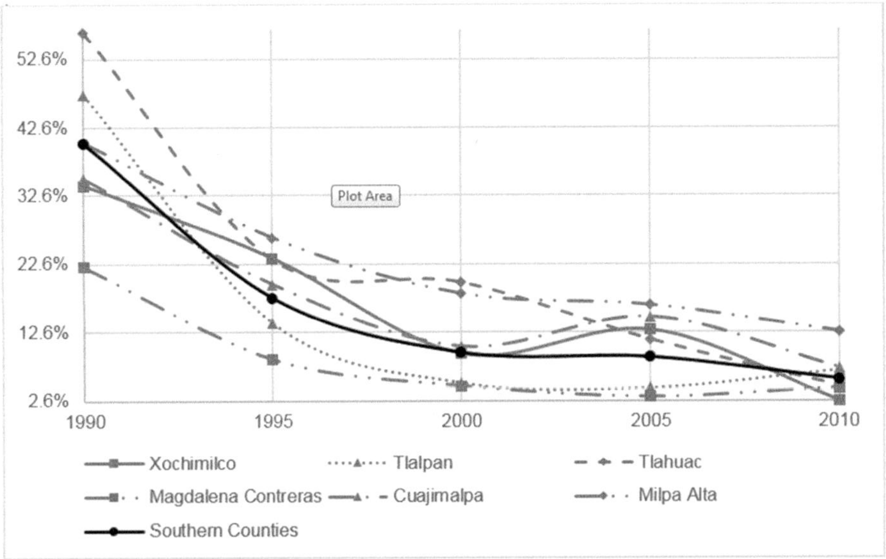

Chart 6.3 Evolution of the population growth rate in Mexico City (1990–2010), Southern Counties
Source: Own elaboration with data from Table 6.1

of tertiary activities at the local level, encouraged by a compact city housing policy. Charts 6.4 and 6.5 show this dynamic by contrasting the growth rates in the groups of counties by functional area, from the first decades of the second half of the twentieth century to the recent years.

In the midst of this population trend in the City, and during the country's transit to a neoliberal development path toward the 1990s, minimum social housing appeared as an "efficient" market response to reduce urban poverty and improve the quality of life. It was the result of an approach based on free market thinking that reduced and replaced the concept of "social housing" with "minimum housing." The housing policy model prioritized the quantitative aspect of housing with emphasis on the mass production of large urban complexes, which implied the reduction of the quality of the physical space, the quality of the materials, and the habitability conditions. This model of housing production shaped the purpose and characteristics of the housing sector and consequently the housing policy of the country, as shown below.

In Mexico, the housing sector plays an important role in the economy because it directly influences economic growth and job creation. According to the "Housing Satellite Account of Mexico 2015," the market and non-market activities accounted in the sector indicate that housing represented 5.9% of

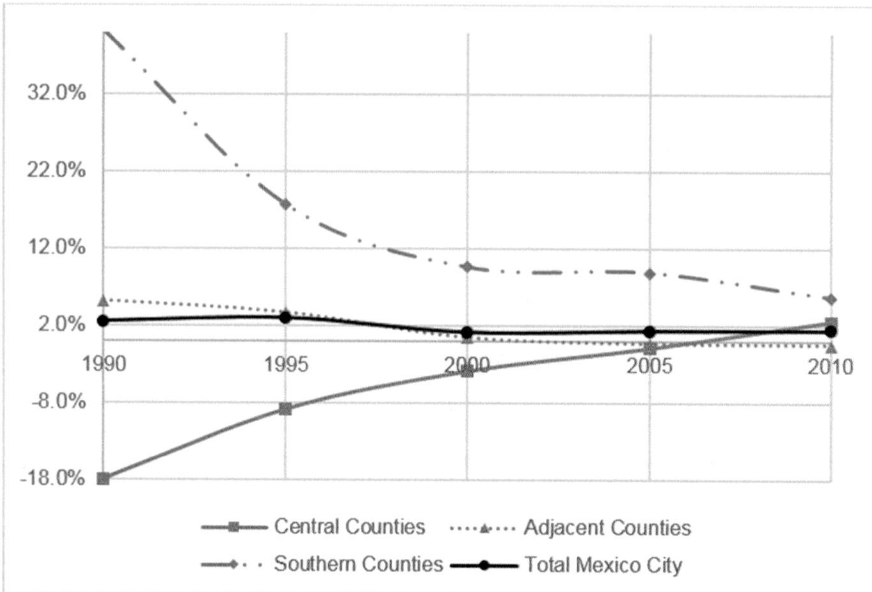

Chart 6.4 Evolution of the population growth rate in Mexico City (1990–2010), Counties or Functional Areas
Source: Own elaboration with data from Table 6.1

GDP. Within this area, 60.2% of this proportion corresponded to the production of goods and services for the market (the construction of the private sector); while 38.7% corresponded to the production for the final use of households (self-construction, expansion, improvement, etc.); and the remaining 1.1% to non-market production (social housing provided by the government). The activities related to this sector constituted 6.8% of the country's occupation.

For INEGI, if the recommendations of the 2008 System of National Accounts are taken into account and rent (imputed) is included within the GDP of housing, it would represent 13.9% of the GDP of the economy in 2015 (INEGI, 2017). According to the latest estimates, the GDP of housing represents 14.1% of the national GDP, and generates more than 3 million jobs, influencing nearly 78 branches of the economy (Caballero, 2016). This fact emphasizes that the housing sector is predominantly private, where the housing production and circulation industry, in addition to having a large, profitable and constantly growing market, does not reflect an important social participation, which leaves access to housing to the self-regulated market.

From the year 2000, the housing sector at national level with its housing policy presents two trends. The first goes from 2000 to 2011, where there

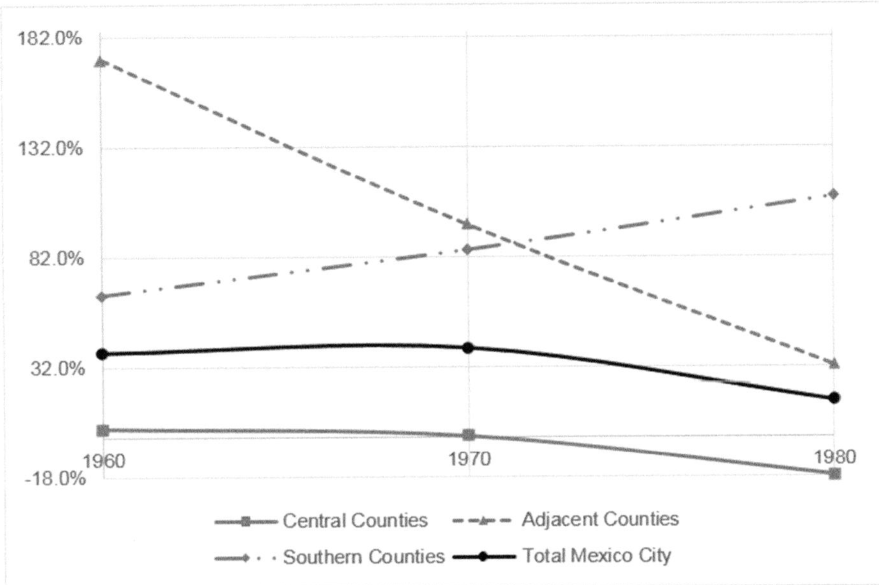

Chart 6.5 Evolution of the population growth rate in Mexico City (1960–1980), Counties or Functional Areas
Source: Own elaboration with data from Graizbord, 1991 and INEGI

is a real estate boom and a significant growth in the sector through policies oriented to the construction of houses in volume. This policy integrated factors such as: "a) Construction of affordable housing to meet the demand and promote the housing sector that was lagging behind, b) Promotion of national housing organizations made up of various institutions, c) Promote the offer of housing in volume and in accordance with the purchasing power of the population" (Caballero, 2016). In this period, with the objective of encouraging the acquisition of new housing, development companies and housing policy established a line of credit oriented to families with incomes below four minimum wages. This channeled efforts to low-cost projects that many developers located in remote areas of the cities to offer low-cost housing, thereby achieving a growing profitability given the increasing volume of inventory that managed to place in the market. This model echoed the overproduction of housing that triggered the crisis in the real estate sector in the United States, with the difference that in Mexico the location of the developments was oriented to the creation of new residential areas in spaces never before urbanized.

The second period is from 2011 onwards. Although paradoxically during the year after the 2008 crisis, there was a GDP growth in the housing sector from 14.8–15.5%, the crisis in this sector was reflected macro-economically from 2011, when the GDP of the sector decreased one percentage point (to 14.5%). Despite the maintenance of the low-cost, high-profit housing production model, the market has registered a steady decline. With the objective of making the market more attractive and incentivizing the growth of the sector, in 2013 the housing policy was oriented to the change in the rules and regulations of the Housing Funds. Far from directing the offer to new localities, the changes had the effect of delaying the construction and financing processes, stagnating the economic acceleration of the sector.

From above, in appearance the housing policy seeks to generate conditions to influence the creation of affordable housing for Mexican families. However, because of the low purchasing power and high prices of "social" housing, housing policy presents a structural limitation derived from an economic policy oriented to deregulation, decoupling from a production project and financing with a strategic development perspective linked to the social sphere.

Regarding social welfare, comparing the development model by import substitution against the neoliberal development path, the economic results are widely contrasting. If during the first model (1935–1982), the purchasing power of minimum wages increased 96.9%; with the country's undertake in the second model, minimum wages lost 69.6% of their purchasing power, that is, they deteriorated to less than a third of those in 1982 (Damián, 2007). In addition, the government reduced the amount of subsidies for basic food products and increased the prices of public goods. Both policies raised the prices of tortilla, beans, milk, meat, among others, as well as water, transportation, electricity, gasoline, fertilizers, etc., affecting the level of consumption of the population and contributing to the increase of poverty (Damián, 2007).

Comparing the evolution of the proportion of poor people in Mexico in the light of the two development models referred above, the tendency toward the reduction in the proportion of the poor reached with the development model by import substitution reversed during neoliberalism. Chart 6.6 shows the comparison of three different poverty lines: that of the Economic Commission for Latin America and the Caribbean (ECLAC), the World Bank (WB), and that of the Normative Basket of Essential Satisfiers (CNSE) integrated by Boltvinik and Hernández Laos. From the period 1981–1984, poverty increased dramatically until 1996 and although it tends to decrease in the next 4 years, compared to the level of the year of departure, high levels of poverty persist.

Boltvinik (2013) describes the dynamics of the rise of neoliberalism in Mexico (1981–1996) as one of the regressive trends of gradual increase in

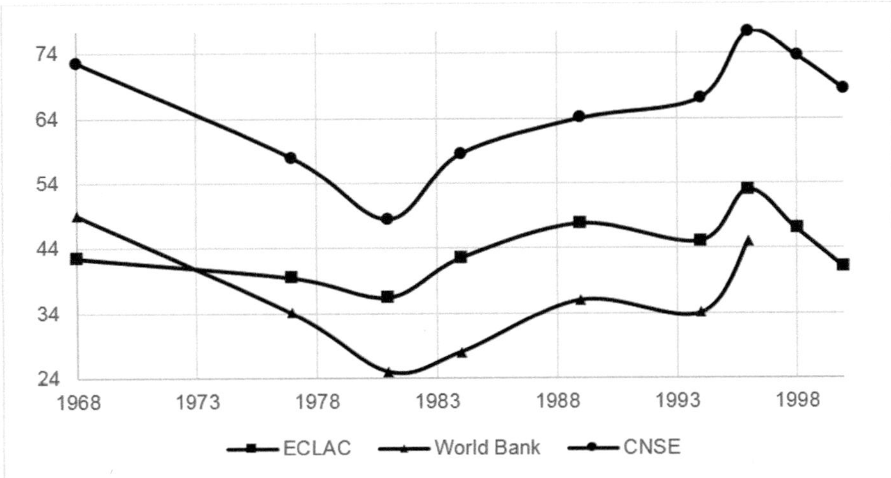

Chart 6.6 Percentage evolution of poverty in Mexico, 1968–2000. Various measurement methods
Source: Boltvinik (2013)

poverty, which reaches levels greater than that of 1968. In the three methods to measure poverty, this increase is notorious: ECLAC registers an increase of 15.5 percentage points in this period, the BM data indicates an increase of 20 percentage points, and with the CNSE method the increase is 28.8 percentage points, which represents a serious setback. There is also a slight and insufficient reduction trend, where the incidence of poverty decreases again without reaching the levels of 1981 and the long-term margin in reference to the initial year of the contrast period is very small. For the year 2000, the percentage of the population in conditions of poverty by the ECLAC method is 41.1% and the CNSE measurement is 68.5%, which represents a reduction of just 1.4 and 4.1 percentage points respectively in 32 years.

From the set of elements provided above, it is highlighted that the country's undertake of a neoliberal development path from the eighties of the twentieth century, and the abandonment of the industrialization strategy by import substitution, exerted a considerable influence on the evolution of population dynamics (including population characteristics and their spatial location in functional areas) of Mexico City. In addition, the free market boom has also shaped the dynamics of housing production and its public policy at the national level, influencing levels of urban poverty and inequality. Under this perspective, in the following section, an exercise will be carried out to measure poverty and inequality induced by lack of housing in Mexico City, as a result of the deindustrialization and depopulation processes promoted by neoliberalism in the country's capital.

6.3 Measuring Poverty and Inequality Induced by Lack of Housing in Mexico City from 2008 to 2015

The notion of poverty induced by lack of housing (PiH) represents a first approach that allows the evaluation of the costs impact associated with covering the population's housing needs as a condition of a poverty situation. It also represents the scope of the national housing policy addressed in the previous section to avoid a situation of vulnerability of the inhabitants who lack their own housing. Camargo and Hurtado (2011), quoting Ruprah (2010), state that the concept of lack of housing–induced poverty refers to "the situation in which total income less housing expenditure (amortization fee or rental expenses, taxes on the property and maintenance costs of the house) is insufficient to acquire the basic family basket "(p. 234). In addition, the same authors quoting Kutty (2005), mention that "subsidies on housing expenditure significantly decrease the probability that households that are slightly above the poverty line will fall into it" (p. 235); This is why the PiH index also represents an indirect way of assessing the effectiveness of public housing policy in a given region.

Under the above conceptual logic, and taking into account that in Mexico City about half of the homes are in a condition of rent, loan, or credit as shown in Chart 6.7; this section presents the calculation method and results of the PiH index in a period from 2008 to 2015 for the City and its 16 counties. This calculation exercise uses the Socioeconomic Conditions Modules (SCM) of the National Household Income Expenditure Surveys of the Mexican Statistics and Geography Institute (INEGI) in its publications of 2008, 2010, 2012, 2014, and 2015. It is important to mention that the design of the SCM survey itself, limits the level of reliability in the generation of results at federative entity. So that the value of the indicator set forth below presents greater validity in the case of Mexico City until 2014, since Statistics and Geography National Institute warns that the 2015 SCM edition is not comparable with previous modules. However, it is possible to consider the index obtained for the City and its counties in 2015, as an indicative or proxy value to appreciate the economic–social phenomenon under study.

Considering the conceptual elements and the statistical limitation set out above, the calculation of the PiH indicator is as follows. The average income of the population in a given region and period, minus the average of the average values of rent, loan, and payment of housing credit, divided by

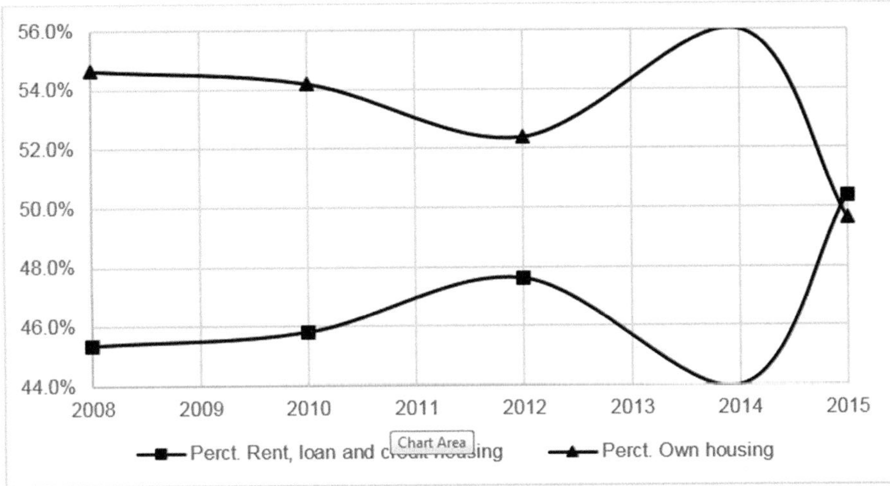

Chart 6.7 Percentage evolution in housing tenure in Mexico City (2008–2015)
Source: Own elaboration with Socioeconomic Conditions Modules data 2008, 2010, 2012, 2014, and 2015.

the value of the food and non-food baskets for the specific calculation year. That is to say:

$$iPiH_{ij} = \frac{\overline{\text{income}_{ij}} - \left(\overline{rent_{ij}}, \overline{loan_{ij}}, \overline{credit_{ij}}\right)}{\text{Wellness Line}_j}$$

Where:
$iPiH_{ij}$ = *Index of Poverty induced by lack of Housing in the county i in the year j*
$\overline{income_{ij}}$ = *Average income of the inhabitants in the county i in the year j*
$\overline{rent_{ij}}$ = *Average cost of housing rent in the county i in the year j*
$\overline{loan_{ij}}$ = *Average cost of housing loan in the county i in the year j*
$\overline{credit_{ij}}$ = *Average cost of housing credit in the county i in the year j*
Wellness $Line_j$ = *Price of food + non food baskets in the year j*

Based on the previous equation, Table 6.2 shows the results of the calculation of the PiH index for the years, as well as for the 16 counties and Mexico City.

From the previous tabulation, the index for Mexico City had a relatively stable behavior during the analysis period. While during the years 2008 and 2010, it maintains a value of 0.9; in year 2012, it presents a significant

Table 6.2 Index of Poverty induced by lack of Housing (PiH) for Mexico City and its 16 counties

County	2008	2010	2012	2014	2015
Alvaro Obregon	–0.7	1.1	1.2	0.8	0.3
Azcapotzalco	1.3	1.3	0.5	1.4	0.6
Benito Juarez	0.9	2.1	0.4	2.3	1.4
Coyoacan	1.7	1.6	2.4	1.1	–0.3
Cuajimalpa	1.2	–0.2	1.9	0.6	1.5
Cuauhtemoc	1.3	0.8	1.6	1.1	1.0
Gustavo A. Madero	1.2	1.0	1.1	0.8	0.7
Iztacalco	1.0	1.0	1.0	0.2	0.6
Iztapalapa	0.9	0.9	1.1	0.8	0.6
Magdalena Contreras	0.7	0.4	0.7	1.6	–0.2
Miguel Hidalgo	1.5	1.2	1.8	–0.2	2.4
Milpa Alta	0.7	0.9	1.6	0.3	1.8
Tlahuac	0.8	0.7	0.9	0.6	0.9
Tlalpan	0.3	1.0	2.3	0.9	0.4
Venustiano Carranza	1.8	1.1	0.9	0.8	1.0
Xochimilco	0.1	0.7	1.3	0.6	0.8
Mexico City	0.9	0.9	1.3	0.8	0.8

Source: Own elaboration with Socioeconomic Conditions Modules data 2008, 2010, 2012, 2014, and 2015.

increase reaching a value of 1.3; and then it decreases to 0.8 in the years 2014 and 2015. This implies that, on average, during 2008 and 2010, an inhabitant of the City who lacked his/her own home, after using his/her disposable income to meet this need through rent, loan, or credit, was able to acquire only 90% of the value of the food and not food baskets; therefore, increasing his/her probability of falling into a condition of poverty. In 2012, the year in which the PiH index acquires a maximum value of 1.3 for the City; the same individual could acquire 100%of the value of the baskets plus an additional 30%, observing a considerable improvement over the previous 2 years, overcoming the Welfare Line and thus escaping the condition of poverty. This situation is reversed in subsequent years (2014 and 2015), where the PiH index has a value of 0.8, from which the individual once he/she meets his/her need for housing, can only access 80% of the value of the food and non-food baskets, once again at risk of falling into poverty.

The previous analysis highlights the importance of housing tenure as a conditioning factor of the population's poverty situation. Particularly of those who satisfy this basic need through rent, loan, or credit, as the percentage of their disposable income destined to cover it, determines the capacity of residual income to acquire all food and non-food baskets or only a percentage of it. Chart 6.8 shows this, graphing the evolution of the ratio

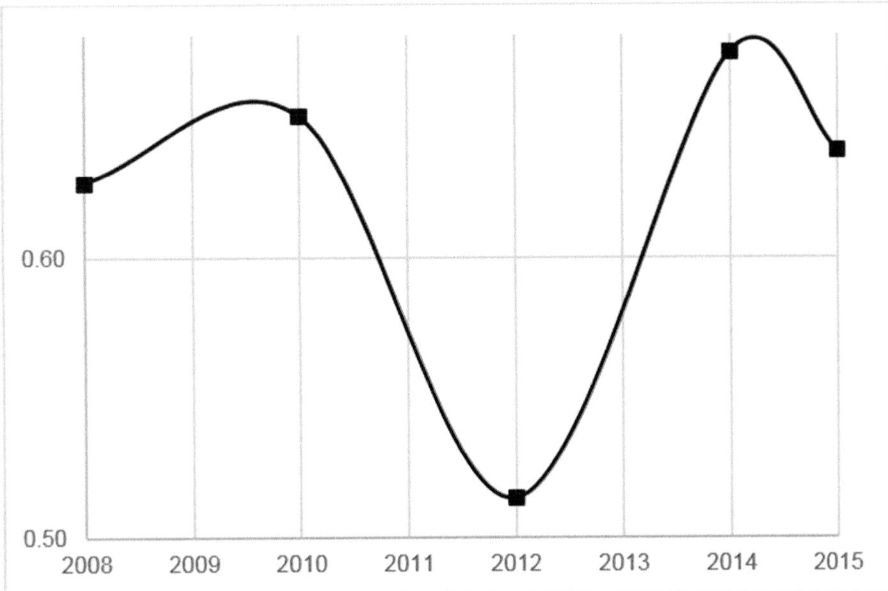

Chart 6.8 Percentage evolution of the quotient: Average of costs associated with housing / Average income in Mexico City (2008–2015)
Source: Own elaboration with Socioeconomic Conditions Modules data 2008, 2010, 2012, 2014, and 2015.

between the average costs associated with housing and the average income in Mexico City. The lowest value in the entire series is in 2012, when the quotient acquires a value of just over 0.5. That is, on average, a person who lacked his/her own home used half of his/her income to solve that need. And with 50% of its remaining disposable income, he/she could acquire the value of food and non-food baskets, which represented 38.5% of its income in that year [see Chart 6.9], having conditions to save up to 11.5% of its income or allocate it to other expenses.

This relative favorable condition does not appear in the City in the years of analysis before and after 2012, when the PiH index takes values of 0.9 and 0.8 respectively. In 2010, on average, the individual who lacks a home allocates a little more than 60% of his/her income to the expenses associated to cover that need [see Chart 6.8] and percentages very close to 40% to acquire the food and non-food baskets [see Chart 6.9], a situation that places him/her in a vulnerable condition. In the second case, for the years 2014 and 2015, adverse conditions tend to be exacerbated compared to the years before 2012. In these years, the individual allocates a percentage close to 70% of his/her

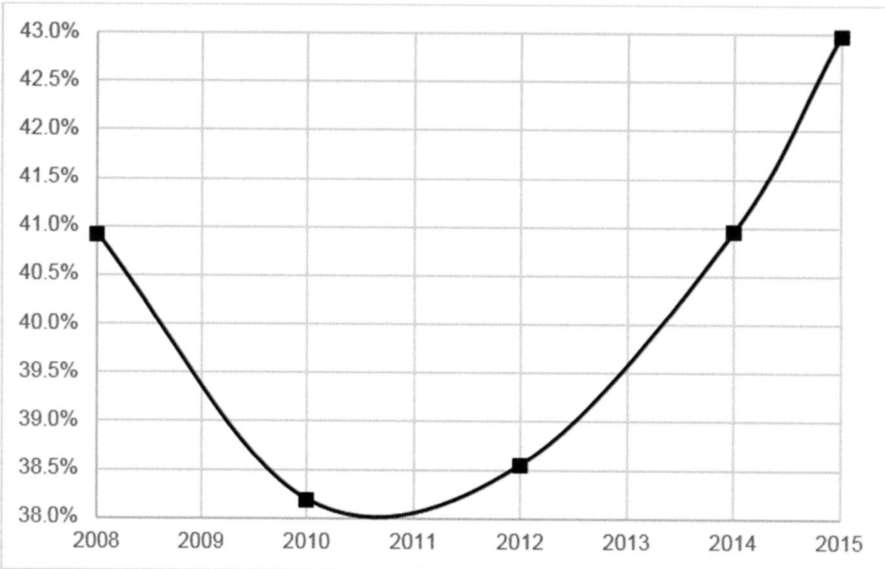

Chart 6.9 Percentage evolution of the quotient: Wellness Line / Average income in Mexico City (2008–2015)
Source: Own elaboration with Socioeconomic Conditions Modules data 2008, 2010, 2012, 2014, and 2015.

income to cover the expenses associated with housing [see Chart 6.8, for 2014 the ratio is 0.67], while the cost of the baskets absorbs just over 40% of his/her income [see Chart 6.9], placing the person in a very probable condition of poverty. As already mentioned, according to the PiH, in 2014 and 2015, the individual can only acquire 80% of food and non-food baskets with residual income after covering the cost of housing.

Based on the index calculation of the population poverty induced by lack of housing between 2008 and 2015, it is possible to obtain a measure of the inequality of the inhabitants of the City based on their homelessness, as well as their space—scalar distribution in the country's capital. This indicator for the 16 counties of the City is the average of the PiH indices calculated in the years of analysis minus the PiH average of Mexico City. From this calculation, the inequality condition of the population induced by lack of housing of a certain county is understood as the differential between the situation of poverty induced by homelessness in that county in relation to the same situation in the City, having the latter as a reference for all counties. Table 6.3 presents the results.

Table 6.3 Average Index of Poverty by lack of Housing (2008–2015) and Inequality index induced by lack of Housing for Mexico City and its 16 counties

County or functional area	PiH Average 2008–2015	Inequality induced by lack of Housing index
Cuauhtemoc	1.2	0.2
Miguel Hidalgo	1.4	0.4
Venustiano Carranza	1.1	0.2
Benito Juarez	1.4	0.5
Central counties	1.3	0.3
Azcapotzalco	1.0	0.1
Gustavo A. Madero	1.0	0.0
Alvaro Obregon	0.5	–0.4
Iztapalapa	0.8	–0.1
Coyoacan	1.3	0.4
Iztacalco	0.8	–0.2
Adjacent counties	0.9	0.0
Xochimilco	0.7	–0.2
Tlalpan	1.0	0.0
Tlahuac	0.8	–0.1
Magdalena Contreras	0.6	–0.3
Cuajimalpa	1.0	0.1
Milpa Alta	1.0	0.1
Southern counties	0.9	-0.1
Total, Mexico City	0.9	0.0

Source: Own elaboration based on Table 6.2

The map of inequality induced by lack of housing in Mexico City [see Map 6.1] shows the Central City, which represents 9.6% of the City's surface, consisting of the Cuauhtemoc, Miguel Hidalgo, Venustiano Carranza, and Benito Juarez counties. This is a relatively homogeneous nucleus in which an average inhabitant of its population has the possibility of covering up to 40 percent more, the cost of food and non-food baskets after satisfying his/her need for housing (by rent, loan, or credit). As mentioned in the first section, this region had a strong process of deindustrialization from the 1990s onwards, as well as an expansion of the tertiary activities of the economy. Thus, causing a double process of expulsion of the less economically favored population, first to the adjacent counties of Central City and gradually toward the metropolitan area of Mexico City. Then pulling in the population with greater economic resources, attracted by the quality of infrastructure, the boom in services, the real estate boom, as well as the public policy implemented between 2000 and 2006, that encouraged repopulation of the heart of the capital of the country.

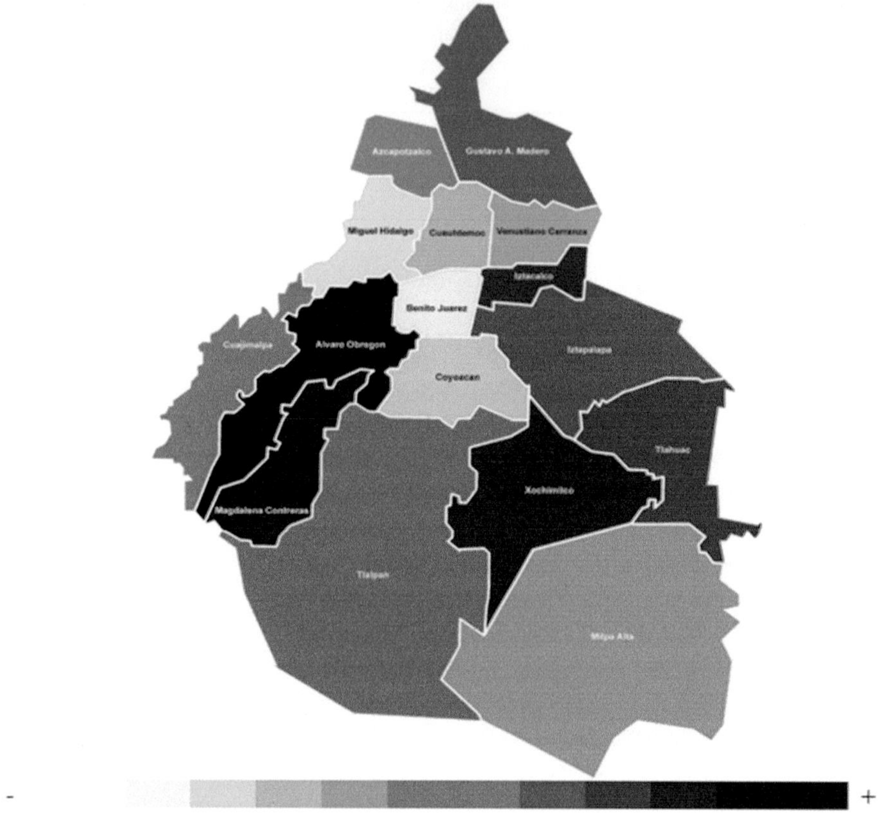

Map 6.1 Inequality induced by lack of Housing in Mexico City and its 16 counties
Source: Own elaboration based on Table 6.3

A heterogeneous layer of counties, where the inequality induced by lack of housing index tends to be more contrasting, surrounds the Central City. In this second layer of the City, which represents 28.8% of its surface, are the counties of Azcapotzalco and Gustavo A. Madero to the north. Their population barely manages to cover their basic needs after satisfying the need for housing. These two counties also underwent a process of deindustrialization from the 1990s; also as a first process of housing the population got expelled from Central City until the year 2000, later their population stagnated in subsequent years [see Chart 6.2]. The county of Coyoacan to the south is mostly oriented to services and the price of land has increased due to real estate boom speculation. In a similar way to Central City, the inhabitants of Coyoacan can acquire up to 30% more than the price of food and non-food baskets, after satisfying their housing needs. On the contrary, the counties of Alvaro Obregon to the west, and Iztacalco and Iztapalapa to the east of the City are the territories of greatest inequality of their inhabitants due to lack of housing in this second layer. In the first case, its inhabitants can only acquire half the value of the basic baskets after satisfying their need for housing; while in the case of the two delegations located to the east, this indicator hardly increases to 80% of the value of the baskets. Alvaro Obregon and Iztapalapa, have presented a consecutive process of depopulation from the nineties; while in Iztacalco there is an increasing trend in the population growth rate from 1990 to 2000, to then stagnate like the previous delegations, all adding to a gradual process of population expelling toward the municipalities of the adjacent states to Mexico City.

Southern counties are a last layer that surrounds the Central City. There are the counties of Cuajimalpa, Magdalena Contreras, Tlalpan, Xochimilco, Tlahuac, and Milpa Alta. They represent 61.6% of the surface of Mexico City and host different areas of natural conservation, having an agricultural tradition as the main economic activity. Therefore, the government has progressively discouraged human settlements in protected areas, promoting a depopulation process of this region and expulsion of its inhabitants to the metropolitan area of the City [see Chart 6.3]. The population of this region present a clear situation of inequality induced to the lack of housing in relation to the inhabitants of Central City and Adjacent Counties. In no case of the Southern Counties, inhabitants can acquire an additional percentage of the basic baskets once they satisfy their need for housing [see Table 6.3], being in a very probable situation of poverty and social vulnerability.

6.4 Conclusion

Neoliberalism as a development modality undertaken by Mexico from the eighties of the last century shaped the production of space and housing in Mexico City. Deindustrialization paved the way to the emergency of tertiary sector activities and the concentration of services in the City. This configured an uneven development pattern, marked by low economic growth in recent decades, real estate speculation, and weak public housing policy. These phenomena let the elites to shape the space, expelling the most disadvantaged population from the heart of the City and rising inequality among the inhabitants due to the lack of housing. Privileged spaces inhabitants in the central counties have a lower probability of falling into poverty, while the population of adjacent and southern counties subsist in conditions of social–economic vulnerability.

That shows the need to rethink the pre-eminence of market as the most efficient agent for social resources distribution. Meanwhile, the return of the

State as a representative of common interests becomes a priority to promote new economic growth and to balance current Mexico City's inequalities of spatial development and housing production. The goal is to minimize its population poverty conditions, and social inequality linked to the lack of housing.

References

1. Boltvinik, J. (2013). Para Reformar la Reforma Social Neoliberal (que ha fracasado) y fundar un auténtico Estado de Bienestar en México. *Estado y Comunes Revista de política y problemas públicos*, 57–94.
2. Boltvinik, J. (2012). Mexico's alleged influence on poverty reduction strategies in Latin America. *Estudios Críticos del Desarrollo, II*(2), 13–40.
3. Caballero, J. (12 de octubre de 2016). Crecimiento del sector vivienda en los últimos años. *El Financiero.*
4. Calva, J. (2004). La economía mexicana en perspectiva. En J. Boltvinik, & A. Damian, *La pobreza en México y el mundo* (pp. 100–132). México: Siglo XXI.
5. Camargo, A. & Hurtado, A. (2011). Vivienda y pobreza: una relación compleja. *Cuadernos de vivienda y urbanismo, 4*(8), 224–246.
6. Damián, A. (2007). Los retos en materia social en México a inicios del siglo XXI. En J. Calva, (Coord.) *Empleo, ingreso y bienestar* (pp. 315–345). México. UNAM.
7. Graizbord, B. (1991). La Población y el empleo en la Zona Metropolitana de la Ciudad . *Cuadernos de Urbanismo*(2), 17–34.
8. INEGI. (2008). *Módulo de Condiciones Socioeconómicas de la ENIGH (MCS - ENIGH) 2008.* Recuperado de: http://www.beta.inegi.org.mx/proyectos/enchogares/modulos/mcs/2008/default.html
9. INEGI. (2010). *Módulo de Condiciones Socioeconómicas de la ENIGH (MCS - ENIGH) 2010.* Recuperado de: http://www.beta.inegi.org.mx/proyectos/enchogares/modulos/mcs/2010/default.html
10. INEGI. (2012). *Módulo de Condiciones Socioeconómicas de la ENIGH (MCS - ENIGH) 2012.* Recuperado de: http://www.beta.inegi.org.mx/proyectos/enchogares/modulos/mcs/2012/default.html
11. INEGI. (2014). *Módulo de Condiciones Socioeconómicas de la ENIGH (MCS - ENIGH) 2014.* Recuperado de: http://www.beta.inegi.org.mx/proyectos/enchogares/modulos/mcs/2014/default.html
12. INEGI. (2015). *Módulo de Condiciones Socioeconómicas 2015.* Recuperado de: http://www.beta.inegi.org.mx/proyectos/enchogares/especiales/mcs/2015/

13. INEGI. (2017). *Cuenta Satélite de la vivienda de México, 2016.* Comunicado de prensa , Instituto Nacional de Estadistica y Geografía, Aguascalientes.
14. Kutty, N. (2005). A new measure of housing affordability: Estimates and analytical results. *Housing Policy Debate, 16*(1), 113 - 142.
15. Ruprah, I. (2010). Do social housing programs increase poverty?: An empirical analysis of shelter induced porverty in Latin America. *OVE Working Papers 0510.* Washington: Inter - American Development Bank - Office of Evaluation and Oversight (OVE).
16. Williams, M. (1997). La distribución del empleo en las delegaciones y los municipios de la ZMCM, 1988-1994. En Coulomb, *Dinámica urbana y procesos sociopolíticos 2 Investigaciones recientes sobre la Ciudad de México* (pp. 55-91). México: Observatorio Urbano de la Ciudad de México.

7

Inequality in the Distribution of Income as at Threat to the Sustainable Development of United Territorial Communities in Ukraine

Ievgen Khlobystov[1], Lidiia Horoshkova[2]

[1]Department of Ecology at the National University "Kyiv-Mohyla Academy" and at the University of Economics and Humanities (Bielsko-Biala, Poland).
[2]Professor of the Department of Business, Management of Organizations and Logistics of Zaporizhzhia National University, Ukraine

ABSTRACT

Current decentralization and reform of the administrative and territorial structure of Ukraine, the issue of overcoming inequalities and disparities of territorial development in Ukraine by building an effective system of financial management for united territorial communities (UTCs) as basis for territories' and country's sustainable development becomes especially relevant.

In the analysis, general scientific methods (analysis and synthesis, induction and deduction) and special methods of phenomena and processes analysis (abstraction, econometric and econometric-mathematical modeling) have been used.

The study analyzes the effectiveness of intergovernmental fiscal relations system and income distribution within the united territorial communities in the context of inequalities and imbalances as a threat to the territories' and country's sustainable development. It has been proved that the current system of fiscal equalization of receipts and expenditure has some disadvantages related to basic and reverse subsidies. It has been proposed to determine the maximum allowable limits of reverse subsidy in relation to income growth rates per one UTC's resident applying the logistical curve. Marginal value of the subsidy has been identified.

Current system of united territorial communities' expenditures, which includes financing of education, culture, health care, social protection, housing and communal services, and infrastructure development has been analyzed in the study. It has been shown that in the current legal framework regarding united territorial communities, there are no effective guarantees of optimal, balanced budget allocation between UTCs settlements—members.

To overcome current disparities in expenditure distribution of a UTC budget, it has been proposed to use economic and mathematical model based on the first-order differential equation, which allows to consider both the dynamics of UTC per capita income and changes in its population.

7.1 Introduction

In September 2015, at the United Nations Summit on Sustainable Development in New York, the Heads of states and governments agreed on Post-2015 Development Agenda defining 17 global Sustainable Development Goals as part of the 70th Anniversary Session of the UN General Assembly. On September 30, 2019, the President of Ukraine signed the Decree "On the Sustainable Development Goals of Ukraine up to 2030." Achieving these goals, adapted to national situation, ensuring real transition of the country and its regions to the model of sustainable development requires application of all the capacity and resources, increasing efficiency of Ukrainian natural resource and socio-economic potential, and overcoming inequalities and disparities in the development of territories as the main objective of public policy.

One of the main tasks of administrative and territorial reform implementation is to remodel relations and powers between administrative territorial units, to set united territorial communities (UTCs) and to grant them master rights to manage their own sustainable development. Decentralization should solve the problem of low standards of local authorities' organization, inefficient management of social development, and regional disparities and inequalities. Therefore, the main task of public policy is to build the framework for the most effective use of territories' natural resource and socio-economic potential as the basis for their balanced development (Bhandari 2020; Bhandari and Shvindina 2019). As there is inequality in territories' basic terms of development, i.e., natural resources, demographic situation, socio-economic development potential, the task of the state amid process of reform is not to deepen the inequality, but to negate the disparities. The tasks could be fulfilled in case of communities' financial capability, maximum power transfer to local authorities, and opportunity to put delegated powers into practice as the basis for sustainable, balanced development of both territories and the country.

It is necessary that the development reforms should be based on scientific principles and recommendations, which allows optimal use of natural resources for the socio-economic development.

7.2 Methods, How Was the Problem Studied?

In the analysis, general-scientific methods (analysis and synthesis, induction and deduction) and special methods of phenomena and processes analysis (abstraction, econometric and econometric-mathematical modeling) have been used.

7.3 Results, What Are the Findings?

The process of decentralization taking place in Ukraine is ambiguous and rather controversial. On the one hand, it is one of the recent positive achievements of Ukrainian authorities; on the other hand, there are numerous problems in its practical implementation.

Therefore, we must say that not all the tasks of the first stage of decentralization (basic reforms) have been successfully completed today. One issue was inequalities and development disparities of territories due to the unresolved or incorrectly resolved problems.

At first glance, due to the amendments to the Budget and Tax Codes, practical steps were taken toward fiscal decentralization, significant changes in intergovernmental budgetary relations and pumping up local budgets due to the redistribution of sources of taxation between different system levels. However, more detailed analysis of the consequences of the implementation of some mechanisms indicates that they impede UTCs' development, threaten their financial capacity, retain inequalities and disparities in territorial development, and make it impossible to ensure country's sustainable development.

According to the regulations of the current Budget Code of Ukraine (Articles 97, 99, 100, 102, 103-2, 103-4, and 108), UTCs budgets have direct intergovernmental fiscal relations with the state budget in the form of: basic subsidy; reverse subsidy; training subvention; medical subvention; subsidies for UTC infrastructure building; subsidies for socio-economic development of different territories; other subventions and grants, if granting and appropriate intergovernmental transfers make sense; and funds from the State

Regional Development Fund (SRDF) (financing for infrastructure projects). Let us provide critical analysis of some of them.

In 2014, the Verkhovna Rada of Ukraine adopted amendments to the Budget Code. The balancing system was replaced by fiscal equalization system, which means horizontal equalization of territories depending on per capita income achieved due to the introduction of basic and reverse subsidy. According to the Budget Code, a basic subsidy is defined as a transfer provided from the state budget to local budgets for horizontal equalization of taxability of territories. A reverse subsidy is funds transferred to the state budget from local budgets for horizontal equalization of territories' taxability (Article 96 of the Budget Code).

Authorities of the Ministry of Regional Development consider that there used to be a balancing system to ensure each level's budgetary liabilities by their expenditure with the income capacity. That is, balancing of local budgets was carried out based on necessary expenditures to support public sector according to the formulaic approach. Expenditures delegated by the state were calculated, then incomes were defined and if expenditures exceeded incomes, the equalization subsidy covered the difference.

Experts of the Ministry of Regional Development emphasize the advantages of current budgetary relations management model. Firstly, there is expenditure-based equalization but not income-based; secondly, equalization is set only for personal income tax (exception regional budgets); it forms the basis for the taxability index of corresponding budget. It is also stressed that this is national tax and other income sources are not involved in the equalization process.

This approach is not without significant financial and economic managerial disadvantages. Firstly, reverse subsidy depends on UTC budget's taxability index, which in turn is determined by personal income tax (PIT). PIT is a major budget income generating source in local budgets' structure. Its share in all local budgets of Ukraine is 56% of total income excluding transfers, and 55% of UTCs in particular. As the share of PIT accumulated by UTCs is quite significant—60%, partial funds withdrawal from this source of income may adversely affect UTCs' financial position.

Secondly, new equalization mechanism does not solve the main problem—revitalization of the activity providing level playing field for national UTCs' financial capacity. Thus, financially capable communities will increase their capacity and transfer some of their incomes to the state budget in the form of a reverse subsidy. At the same time, disadvantaged communities will receive funds from the state budget in the form of a basic subsidy. This mechanism does not support financial sustainability of the vast majority of UTCs but pose a threat to the financial position of UTCs able to increase their income independently.

According to the information about local budgets' creditworthiness and stability published by the Ministry of Regional Development (based on the analysis of territorial communities' financial indicators), basic subsidy negatively affects UTCs' sustainable development if its ratio to own revenues is more than 50% (calculated by division of basic subsidy to total own incomes and basic subsidy). Basic subsidy within 20-30% is considered to be uncritical for a budget. However, according to the analysis, there have been any procedures initiated by the State regarding steadily subsidized UTCs during the reform period.

Moreover, similar standards for the reverse subsidy (at least formal ones) have not been developed. This is a significant failure as the reverse subsidy directly affects stability of local budgets and financial capacity of UTCs. We applied economic and mathematical modeling and proposed a corresponding model.

It was proposed to determine allowable limits of reverse subsidy correlated to UTC per capita income growth rate. It was proved that to describe the dynamics of UTC income, it is expedient to not use the exponential law (hard model), but the logistic model (soft model). It was offered to determine safe limits of UTCs' financial position's adjustment with the help of the logistic model ensuring sustainable development. Conditions for the external impact on UTC system in the form of reverse subsidy were shaped. Safe limits for its amounts' adjustments were defined as well. It was proved that 50% of UTC income is the critical amount of the reverse subsidy. The estimated allowable amount of the reverse subsidy is no more than 25% of UTC income [1].

The analysis of reverse subsidy to UTC income ratio over time was made to prove the conclusions regarding the deterring effect of reverse subsidies and current inequality of UTCs' development in Ukraine. It showed that Honcharivska UTC (Chernihiv region); Verbkivska UTC (Dnipropetrovsk region), and Bogdanivska UTC (Dnipropetrovsk region) occupied first places during 2016–2018s. Besides, it is noteworthy that reverse subsidies for these communities increased from 2016 to 2018: in Verbkivska UTC (Dnipropetrovsk region)—from 15.92% to 19.9% and 24.9%; in Bogdanivska UTC (Dnipropetrovsk region)—from 13.9% to 19.1% and 23.4%. In Honcharivska UTC (Chernihiv region)—from 25.8% in 2017, from 25.8% to 29.3% in 2018. Honcharivska UTC was set in 2016 so its budget in 2016 had no subsidies.

Average income per capita growth rate analysis for these communities during 2015–2017s demonstrated the following results: Verbkivska

UTC—946.22% (2016/2015), 25.98% (2017/2016); Bogdanivska UTC—632.95% (2016/2015), 24.50% (2017/2016).

Thus, the comparison of own per capita income analysis' results for these communities during 2015–2017s showed that the growth rate of community income was distributed as follows: Verbkivska UTC (Dnipropetrovsk region)—946.22% (2016/2015), 25.98% (2017/2016); Bogdanivska UTC (Dnipropetrovsk region)—632.95% (2016/2015), 24.50% (2017/2016). That is, exponential income growth was only in the first year (2015). In 2016, income growth rates declined substantially, similar dynamics were observed in the first half of 2018 and came closer to logistical pattern. Meantime, during 2016–2017, the reverse subsidy expanded. Reverse subsidies almost reached 25% level calculated as safe for UTCs. Their further growth threatens gradual crisis for UTCs.

We also conducted analysis of basic granting and the reverse subsidy withdrawal based on the results of the first half of 2018, by regions of the country (Table 7.1).

The results indicate that among the regions of Ukraine there are those with a significant part of UTCs receiving a basic subsidy, which has reached or on the threshold. Among them are Lviv, Volyn, Chernivtsi, Ivano-Frankivsk, Ternopil, and Rivno regions. Besides, it was found out that most communities receive a basic subsidy of 30–50% in Lviv, Ivano-Frankivsk, Ternopil, and Rivno regions. At the same time, it should be noted that the level of reverse subsidies is quite high—from 18.8% to 29.3% in Poltava, Dnipropetrovsk, Chernihiv, Mykolaiv, Sumy, and Kyiv regions. The level of basic subsidy is quite low here—from 21.8% to 34.7%.

In our opinion, results obtained prove ineffectiveness of the current mechanism of intergovernmental budgetary regulation as it forms the basis for inequalities of UTCs development characteristics, which may threaten territorial and national sustainable development. As we can see, current mechanism forms the framework when some regions of the country make no effort to work without basic subsidies. Meanwhile, regions having stable indicators of their economic development find out that reverse subsidy slows down further development of their potential, as it withdraws part of funds from UTC budget.

Thus, implementation of reverse and basic subsidies creates inequality in UTCs development and their ability to ensure financial capacity and self-sufficiency as one of the main tasks of decentralization reforms.

Other types of subsidies are not without their disadvantages too, as they also lead to social inequality in UTCs development.

UTCs have one exclusive subsidy type—for UTC infrastructure building, imposed in 2016 in accordance with the resolution of the Cabinet of

Table 7.1 Quantitative indices of basic granting and the reverse subsidy withdrawal based on the results of the first half of 2018

Region	Number of UTCs on 01.07.2018.	Number of UTCs with basic granting	Maximum basic subsidy, %	Number of UTCs withdrawing reverse subsidy	Maximum reverse subsidy, %
Dnipropetrovsk	56	38	34,7	9	24,9
Zhytomyr	45	36	39,8	7	17,5
Volyn	40	33	61,7	7	23,5
Ternopil	40	36	50,1	2	14,7
Poltava	39	17	22,4	14	19,7
Khmelnytskyi	39	34	41,7	4	15,5
Khmel'nychchyna					
Khmelnytskyi					
Chernihiv	37	21	22,6	8	29,3
Zaporizhzhia	36	32	45,2	2	8,7
Lviv	35	31	55,8	4	6,0
Vinnytsia	34	22	41,0	9	11,4
Mykolaiv	28	22	36,9	5	20,8
Sumy	28	18	26,4	4	18,8
Kherson	26	24	42,7	1	4,5
Cherkasy	26	18	21,8	4	21,2
Chernivtsi	26	25	53,0	1	4,3
Odessa	25	20	37,7	3	4,4
Rivno	25	22	49,3	2	7,8
Ivano-Frankivsk	23	22	59,0	1	9,6
Kirovohrad	13	4	19,8	6	15,6
Kharkiv	12	9	18,0	2	18,7
Donetsk	9	3	48,0	4	8,9
Kyiv	9	4	8,2	4	21,6
Luhansk	8	6	45,7	1	8,1
Zakarpattia	6	3	43,6	3	0,8

Ministers of Ukraine "Some issues of granting a subsidy from the state budget to local budgets for the formation of infrastructure of the united territorial communities" of 16.03.2016, No. 200. It covers 10 main forms of funding: a) development of project, urban planning, and design documentation; b) quality improvement of administrative services, including setting and modernization of Administrative Service Centres (ASCs), of acquisition of equipment and software; c) formation of modern systems of community management organization, i.e., communication networks, databases, public warning systems; d) reconstruction, reorganization, and repurposing of budgetary institutions buildings applying obligatory energy efficient technologies; e) new construction, reconstruction, and capital improvement of streets, roads, bridges, and communal property transfers; f) acquisition of school transport, special purpose vehicles and their accessories for public utilities, fire and special rescue equipment and fire-fighting equipment, specialized medical transport for healthcare institutions; g) new construction, reconstruction, and capital improvement of water supply and drainage system facilities, waste management units and remediation of landfills, etc; h) other activities with public utilities to ensure adequate security and civil defence level; i) meeting budgetary accounts payable according to the legal procedure under the program "State budget subsidy to local budgets to build UTCs infrastructure"; j) construction, reconstruction, repair and maintenance of local roads, streets and roads of communal property in settlements, as well as repairs of streets and roads that are constituent roads of national significance (co-financing on contractual basis).

The amount of subsidy to build UTCs infrastructure is specified one, which depends on UTC area and its rural population. Therefore, inequalities may arise from both rural and urban settlements in some UTCs. The last have to seek resources for infrastructure development from their own sources. This situation is unfair, as rural residents cannot be restricted to use infrastructure of cities where they, for example, work. Their development funding is not covered by this subsidy.

Other two subsidies—for the activities, which support socio-economic development of individual territories and funds of the State Fund of Regional Development—depend on the decisions of special Commission of the corresponding ministry, which is the main manager of these funds. This indicates subjectivism of the granting process, which can also lead to inequality in their distribution between UTCs.

Therefore, these shortcomings of state financing mechanisms of UTCs development should be improved because they cause inequalities in UTCs development in Ukraine and prevent sustainable development of the country.

These disadvantages can be attributed to UTCs inequalities within districts and country as a whole. There are also some unresolved issues regarding allocation of resources between settlements within one UTC.

Resource management is the basis for complex systems performance. To analyze and evaluate the quality of management, they use the category of production and resource capacity, which characterizes the maximum capacity of the accumulated and prepared for processing natural, material, technical, labor, financial and information resources to meet the needs of society and individual citizens.

However, there is the law of scarce resources. According to this law, available resources do not always fully satisfy all the needs and desires of people.

At the state level, society must choose products for manufacturing, considering opportunities and available resources to balance supply and demand. Besides, there are vital goods and services that are of social importance and the state must ensure their supply to all citizens without exception.

There is similar issue of resource supply at the mezolevel. Each territory is characterized by its specific production and resource potential, and thus by its industrial specialization. At the administrative-territorial level, as well as at the state level, there is a need to supply population with vital goods and services, regardless of abundance or scarcity of certain resources.

Resource potential (ownership right and right of disposition of resources) and the level of rational, effective management are the main determinants of building capable UTCs amid current decentralization.

The ability to manage a community depends essentially on rational and efficient managerial processes impact tangible and intangible benefits making.

It was emphasized in "The methodology of capable communities formation" [2] that: capable territorial community is a territorial community of villages (settlements, cities), which, as a result of voluntary association, are able to provide, on their own or through appropriate local governments, can maintain adequate level of service supply, in particular in the fields of education, culture, health, social protection, utilities, taking into account human resources, financial support and infrastructure development of corresponding administrative-territorial unit.

Our analysis proves that there is no effective guarantee of optimal, balanced distribution of resources among UTCs settlements—members in the legal framework for united territorial communities. This problem arises first in UTCs, which consist of settlements with different population and different per capita income [3]. Thus, there are some shortcomings.

The first corresponds to the fact that as single council of UTCs is formed by equal and direct elections, the number of representatives of each settlement depends on its population. That is, settlements as UTC centres may have decision-making advantages by the population size.

The second problem is that per capita income does not always depend on population size for settlements; more priority is given to natural resources abundance, valuable land or big profitable business. It can also lead to uneven use of funds, for the development of settlements.

The third problem is that fair, balanced expenditures, and subsidies distribution is possible if there is reliable information on the number of UTCs residents. The information is practically absent, since UTC formation is preceded by the elections announced by the Central Election Commission (CEC), based on information about resident's official registration in settlements. Practice shows that the place of registration does not always coincide with the place of residence. In addition, there is a pendulum labor migration in the process of job search between settlements. That is, there are those who work in another locality leaving every morning and returning in the evening. There are those who work during a week, there are also seasonal jobs. Thus, the number of settlements' residents fluctuates. In this case, the traditional approach is that the distribution of necessary expenditures between settlements will not be effective, since it is based on the place of registration without taking into account the temporary migration. In turn, settlements accumulating additional workforce every morning must be provided with appropriate infrastructure (transport, catering, etc.). Transport infrastructure between settlements of such communities also needs additional attention.

We believe that there is another problem with the inefficiency of allocating both of UTCs budget expenditures to individual settlements and consideration of external migration, above all—labor migration outside the country. In regions with significant external labor migration (emigration), the actual population size does not always coincide with that considered by CEC when setting a UTC. It is also not possible to consider the current system of expenditures and other financial resources (subsidies) distribution from the State budget.

Therefore, there is a need to build a mechanism for fair distribution of UTCs budget.

There is experience in solving this problem. Thus, in the EU, there is a principle of harmonious development, enshrined in the Treaty on the Functioning of the EU. It means distribution of the majority of budget to less developed territories. This approach, in our opinion, poses threat of these settlements' affiliation with UTCs or the emergence of communities that would prefer subsidized development instead of creating conditions for their own financial capacity.

Another opposite approach is economic liberalism. It provides dependency of budget expenditures on each participant contribution. The negative experience with this approach is already known in the USA. Due to the threat of social injustice, there are already more and more municipalities separating the rest of the territory, along with the richest districts and, accordingly, their incomes, leaving a gap in the financing of total expenditures. However, this approach will encourage communities to seek new opportunities to earn money.

That is why, it is advisable to use differential approach for the formation of budget expenditures, considering their peculiarities based on social justice principles for UTCs' population.

7.4 Discussion

We consider that it is possible to consider the mentioned shortcomings in distribution of resources between UTCs settlements applying equation of the economic growth model by R. Solow. It looks as follows:

$$\dot{k} = lf(k) - (\alpha + \beta)k \tag{7.1}$$

Let us find the solution of R. Solow model equation for Cobb–Douglas macroeconomic production function:

$$F(K,L) = K^a L^{1-a}, 0 < a < 1$$

Let $Y = F(K,L)$ be total UTC income, that is own income, infrastructure subsidy and basic / reverse subsidy.

F—homogeneous first-order production function described by equation:

$F(tK, tL) = tF(K, L)$

where K—UTC income;

L—UTC population.

Let us introduce index $k = K/L$, which is equal to UTC own income to UTC total population ratio, so we have index of own income per capita for UTC.

Then capital productivity is:

$$f(k) = \frac{F(K,L)}{L} = F(k,l). \tag{7.2}$$

We assume that we have a natural increase in UTC population over period of time:

$$\dot{L} = \alpha L \tag{7.3}$$

where α—coefficient (growth rate) of UTC population.

UTC investments (capital expenditures) are used to increase own resources (income) and depreciation of fixed capital, i.e.,

$I = \dot{K} + \beta K$, where β—depreciation rate (share of capital expenditures).

Then if l—rate of investment, then

$I = lY = \dot{K} + \beta K$,

or

$$\dot{K} = lF(K, L) - \beta K. \tag{7.4}$$

According to the own income per capita definition k, we have $lnk = lnK - lnL$. We differentiate this equation by t, and obtain:

$$\frac{\dot{k}}{k} = \frac{\dot{K}}{K} - \frac{\dot{L}}{L}.$$

We substitute into the last ratio equations (7.3) and (7.4) and obtain the equation of unknown function k having the form (7.2), where function $f(k)$ is defined by formula (7.1). This first-order nonlinear differential equation to own income per capita has simple economic interpretation: net own income increment is the difference between gross own income and steady-state own income.

Equations of R. Solow model for Cobb–Douglas production function take into account $F(K, L) = \dfrac{K^a L^{1-a}}{L} = (\dfrac{K}{L})^a = k$, are as follows:

$$\dot{k} = lk^a - (\alpha + \beta)k, 0 < a < 1. \tag{7.5}$$

We integrate the Bernoulli equation by the substitution method.

Let $k = uv$. then

$$\dot{k} = \dot{u}v + \dot{v}u$$

equation (7.5) we set as:

$$\dot{u}v + \dot{v}u = l(uv)^a - (\alpha + \beta)uv$$

or

$$\dot{u}v + u(\dot{v} + (\alpha+\beta)v) = lu^a v^a. \tag{7.6}$$

Taking into account that one of the unknown functions, such as v, can be arbitrarily chosen (because only derivative uv must meet original equation), we take any partial equation solutions for v

$\dot{v} + (\alpha+\beta)v) = 0$, which turns to zero coefficient of u in equation (7.6). Obtain:

$$\frac{dv}{dt} = -(\alpha+\beta)v.$$

After integration, we get: $\ln|v| = -(\alpha+\beta)v$ or $v = e^{-(\alpha+\beta)t}$ (we do not introduce continuous integration because only a partial solution of the auxiliary equation is required). To calculate u we have equations

$$\dot{u}v = lu^a v^a$$

or

$$\dot{u}e^{-(\alpha+\beta)t} = lu^a e^{-a(\alpha+\beta)t}.$$

We divide variables and obtain:

$$\frac{du}{u^a} = le^{(\alpha+\beta)(1-a)t}dt,$$

then

$$\frac{u^{-a+1}}{1-a} = l\frac{1}{(1+a)(\alpha+\beta)}e^{(\alpha+\beta)(1-a)t} + \frac{C}{1-a}$$

or

$$u = \left(\frac{l}{(\alpha+\beta)}e^{(\alpha+\beta)(1-a)t} + C\right)^{\frac{1}{1-a}}.$$

Then

$$k = \left(\frac{l}{(\alpha+\beta)}e^{(\alpha+\beta)(1-a)t} + C\right)^{\frac{1}{1-a}} e^{-(\alpha+\beta)t} =$$

$$= \left(\frac{l}{(\alpha+\beta)} + Ce^{-(1-a)(\alpha+\beta)t}\right)^{\frac{1}{1-a}}.$$

When t→∝, own income per capita is:

$$k \rightarrow \left(\frac{l}{(\alpha+\beta)}\right)^{(1-a)^{-1}}.$$

Thus, we have proposed the model that simultaneously takes into account the capital expenditures growth rate and changes in population, if UTC per capita income and gross income rise. If necessary, it is advisable to build this function for average UTC indices, and then to determine the expenditures for each UTC's settlement taking into account its change in population in case its per capita income meets UTC average.

Similar calculations can be further made within the region to assess the level of financial autonomy of both UTC and region in general. Comparison of regional indicators (in terms of districts) will allow estimating the level of sustainable development of a territory (region). It will also make it possible to compare these parameters between different regions.

7.5 Conclusion

The paper analyzes the effectiveness of intergovernmental relations and income distribution system within united territorial communities in the context of inequalities and imbalances as a threat to sustainable development of territories and the country.

The analysis, fiscal equalization system introduced in 2014 instead of income and expenditures balancing system made it possible to conclude that the current system is also not without drawbacks. The main one is a reverse subsidy determined by personal income tax that negatively affects UTCs' financial position, since PIT is the main source of community income. It has been proposed to determine the maximum allowable limits of reverse subsidy correlated to per capita income growth rate of UTCs. A so-called soft model, in the form of a logistic curve, has been proposed to describe the dynamics of UTCs' own income. Secure limits of external impact, such as a reverse subsidy, on UTCs' financial position have been determined.

It has been proved that current mechanism of intergovernmental budget regulation causes inequality of development of UTCs because funds in the form of reverse subsidies are being withdrawn from self-sufficient, financially capable communities. Nevertheless, there are UTCs in Ukraine that have been receiving a basic subsidy for a long time, which does not encourage them to support their self-sufficiency. Implementation of this mechanism creates imbalances in regional development funding.

Current system of united territorial communities' expenditures, which includes financing of education, culture, health care, social protection, utilities, and infrastructure development have been analyzed in the study. It has been shown that in the current legal framework regarding united territorial communities, there are no efficient guarantees of optimal, balanced budget distribution between UTCs settlements—members.

Distribution issues have been identified, namely: UTCs include settlements with different population size, different per capita income; there is volatility of population caused by migration processes of different duration. Therefore, the need to build mechanism for efficient UTCs budget distribution based on the real needs and capabilities of communities has been proved. Foreign experience does not have effective mechanisms for budgets expenditures distribution, especially when we consider Ukrainian specifics. Different UTCs can be dramatically different in terms of their capability and potential, even in one administrative region. Therefore, distribution of communities' financial resources should be as efficient as possible and independent of short-sighted decision-making.

To overcome existing disparities in UTCs budget expenditures distribution, it has been proposed to apply economic and mathematical model based on the first-order differential equation, which allows to consider simultaneously the dynamics of UTC per capita income and its population change.

Application of the proposed models has practical significance as it provides opportunity to increase public policy efficiency to affect inequalities and imbalances as a principle for harmonious development of the country and its territories.

References

1. Bhandari, Medani P. (2020) Second Edition- Green Web-II: Standards and Perspectives from the IUCN, Policy Development in Environment Conservation Domain with reference to India, Pakistan, Nepal, and Bangladesh, River Publishers, Denmark / the Netherlands. ISBN: 9788770221924 e-ISBN: 9788770221917
2. Bhandari, Medani P. (2020), Getting the Climate Science Facts Right: The Role of the IPCC, River Publishers, Denmark / the Netherlands-ISBN: 9788770221863 e-ISBN: 9788770221856
3. Bhandari, Medani P. and Shvindina Hanna (2019) Reducing Inequalities Towards Sustainable Development Goals: Multilevel Approach, River Publishers, Denmark / the Netherlands- ISBN: Print: 978-87-7022-126-9 E-book: 978-87-7022-125-2

4. Horoshkova L., Volkov V., Kapranova L., Комelina A. (2018). The Reverse Subsidy`s Impact of United Territorial Community`s Budget Generations. *International Journal of Engineering & Technology*. № 7(4.8). C. 539-543.
5. Government of Ukriaine (2015). On approval of the Methodology for the formation of capable territorial communities: Resolution of the Cabinet of Ministers of Ukraine dated April 8, 2015, No. 214.). URL: http://zakon2.rada.gov.u. (date of treatment 04.02.2018). [In Ukrainian] [Про затвердження Методики формування спроможних територіальних громад: Постанова Кабінету Міністрів України від 8 квітня 2015 р. № 214 зі змінами і доповненнями. URL: http://zakon2.rada.gov.u]
6. Horoshkova L., Khlobystov Ie., Filipishyna L., Bikulov D. (2019). Distribution of Local Expenditure for Sustainable Development of United Territorial Communities (economic and mathematical approach). Advances in Economics, Business and Management Research. Volume 99. P.20-25. https://www.atlantis-press.com/proceedings/mdsmes-19/125919182

8

A Moral and Pragmatic Analysis of Debt and North–South Economic Inequality

George Atisa

The University of Texas Rio Grande Valley, 1201 W University Dr.
Edinburg Texas, USA, Email: george.atisa@utrgv.edu

ABSTRACT

The North–South inequality hurts both high- and low-income countries in many respects, both directly and indirectly. At the same time, workers from low-income countries work just as hard as those from high-income countries but are often held in abject poverty by circumstances not of their own making. The question the study tries to find answers for is: why do hard-working individuals in both the North and the South bring home different levels of income, yielding big differences in per capita income globally and what can be done about it? Current pro-poor growth policies that include handouts, loans or relocating industries to the South cannot rid the world of inequality. The study argues that rich countries can and should prevent poor countries from sinking deeper into poverty if that does not call them to sacrifice anything of comparable moral importance. Data on GDP per capita and natural resources within countries is used to show, and explain two forms of inequality; income levels and natural resources and that every region has something valuable to offer toward maintaining the global economy. The South is endowed with surplus natural resources that the North needs and the North has high incomes that can be used to support development needs in the South. Institutional weaknesses aside, the most feasible framework to reducing global inequality is a pragmatic and moral approach to valuing the natural resources from the South on the same scale as financial development assistance from the North.

8.1 Introduction

This chapter examines global inequality through a moral and pragmatic lens. Moral and pragmatic values are a necessary consideration because global inequality in a globally interdependent world equally hurts both high- and low-income countries in many respects, both directly and indirectly. Following the moral compass while adopting a pragmatic approach means taking a broader view to finding feasible and durable solutions to global inequality. The interdependence can be explained using income and resource inequality. Countries in the North have very high incomes but great deficits in natural resources, while many countries in the South have high surpluses in natural resources but very low incomes. Therefore, no one country or region can claim to have everything that is required to support economic development within its borders. It is most likely that global interdependence will continue to rise, and for this reason, reducing global inequality is a practical goal, as every country or region has great value in a globalizing world.

However, inequality also hurts low-income countries beyond mere material deprivation and touches on the core aspect of human values. Workers from low-income countries work hard but are often held in abject poverty by circumstances not of their own making (Atisa, 2020). Workers from high-income countries also work hard and are able to get ahead. Why do hard-working individuals in both the North and the South bring home different levels of income, yielding such different per capita income levels and what can be done about it? This is a difficult question to answer; if there were a direct answer, the problems leading to economic inequality between the North and the South would have already been solved.

The countries and people at the lower end of inequality face severe poverty, other social ills and crimes that are likely to continue unless poverty is eradicated; therefore, inequality should be seen as a moral issue and not just a business outcome. This study is about the inequality between a deprived society and a society that has more than enough prosperity. The World Bank (2016) argues that poverty eradication is one of the greatest challenges facing many countries, and the ways to eradicate poverty are clearly understood but very difficult to implement. Current approaches to reducing global economic inequality are grants, loans, trade and investments, which have been presented as the distribution of international resources beyond national boundaries (Schaffartzik *et al.* 2019; Bhandari 2020; Bhandari and Shvindina 2019) but have not brought about the desired outcomes. Since income seems to be the standard measure of development outcomes based on the flow of resources, the concentration of resources and incomes levels must be addressed if global inequality is to be reduced.

8.2 Implications of Economic Inequality and Natural Resource Consumption

While global economic inequality is perceived as responsible for global poverty, its damage to humanity goes beyond the realm of poverty. Very low consumption of natural resources is strongly connected to low income and poverty. Wilkinson and Pickett (2014) offer a humbling perspective of inequality using the logic of animal dominance and subordination, superiority, and inferiority as one of the ways that humans equate outward wealth with a person's inner worth and how humans see and feel toward each other. Toward the end of the 20th century, there was a surge in the global consumption of natural resources that led to high inequality between countries (Schaffartzik *et al.,* 2014). The countries with high per capita income and some of the emerging economies, such as China, consumed almost twice as much global resources as the other countries, while the countries with low per capita income subsisted on approximately one-fourth as much (Schaffartzik *et al.* 2019). The high consumption of natural resources by high-income countries is largely ignored in the inequality equation and needs to be part of the global inequality discussion.

Identification of the quantity of global resource appropriations that largely goes to high-income countries to fill their resource biocapacity deficits and support their high per capita income demand can show the extent to which low-income countries are disadvantaged in the global economy. Even though higher efficiency is often touted as a way to reduce waste, high incomes that lead to higher consumption are often accompanied by increased use of resources (Duro and Teixidó-Figueras 2013). Moreover, GDP per capita as a measure of economic performance does not account for the sources and depletion of global resources; rather, it merely hides the real use of resources that brings about overall income levels (Duro and Teixidó-Figueras 2013). This means that high-income countries are directly consuming natural resources from other countries with surpluses without giving credit to such sources.

The contribution of this study is that it tries to connect high income with the unequal use of global resources and makes this a moral justification for rich countries to help poor countries. High incomes and, by extension, economic inequality is generated from the higher consumption of global resources. The primary driver of income, in addition to domestic ingenuity, is the availability of natural resources, which must be seen together; ingenuity on its own cannot bring about the robust development seen in high-income countries. The next section provides a review of the literature and then the methods. Findings are then provided, followed by discussions and finally, conclusions.

8.3 Literature Review

8.3.1 Inequality and Economic Development

In the North–South nexus, the foundation for economic development is perceived to be infrastructure development (World Bank 2016). Many poor countries in the South do not have sufficient domestic revenue sources to build a network of transportation and communication infrastructure (Fuss et al. 2016) to start to get their people out of poverty. The infrastructure development approach that has been presented as helping the national development of a country does not go far enough and is simply designed to facilitate material flows to the North (Biesebroeck and Sturgeon 2010). Through these infrastructure networks, high-income countries are able to access distant natural resources that are not available locally. The infrastructure development mantra is presented as helping resource-rich countries, while in reality, providing access to powerful countries to exploit natural resources in poor countries is morally not right. How then can economic development that helps the resource-rich countries in the South be contextualized? Answers to this question are explored below.

On the basis of moral values, Singer (1972) highlights one key principle that helps find answers to the question raised above: if the rich countries "have the power to prevent something bad from happening without sacrificing anything of comparable moral importance, they ought, morally, to do it." This principle is further explained as preventing something bad and not promoting what is good. This study takes a different view on promoting what is good. In regard to the sharing of natural resources, the value that these resources bring to the North should be estimated and explained as materials of high value that the South contributes to the North so that the North can offer help to the South via financing on the same scale.

This argument is further strengthened by the view that if the opportunity costs of assisting poor countries do not hurt the North economically, providing such assistance is also morally the right thing to do. Working toward reducing global inequality would require pragmatic approaches that use moral values to move global cooperation away from strategies that advance instrumental ends and the exploitation of poor countries by rich countries. Therefore, in a globally interdependent world where some countries have per capita income more than 100 times greater than that of countries classified as languishing in severe poverty (Kaya and Keba 2011), it is only fair that development assistance be considered more holistically, beyond the business framework. From a global resources' perspective, just as the South is supplying the North with much-needed resources to develop, it is also morally right for the North to help the South develop by providing financing.

It is important to re-examine the existing development frameworks, such as loans with very high interest, that have continued to sink poor countries into debt and reduced the capacities of many poor countries to service their domestic programs.

There is also a false impression emerging from a situation where the North has relocated its manufacturing industries to the South, making countries in the South look like they have industrialized (Revenga and Dooley 2019). This has been one attempt to justify the idea that it is not necessary to do anything to address inequality because from an industrial lens, inequality has been narrowing (Arrighi et al. 2003). The industrialization and income gap between the developed North and the developing countries of the South send contradictory signals of succcss. Thc industrialization gap has been narrow ing, but this has not translated into the narrowing of the income gap, which is essentially the aim of industrialization (Arrighi et al. 2003). The industrial gap has been closing in part because of the relocation of industrial enterprises from the North to the South, making it appear that the South is industrializing and catching up with the North. These relocated industries have created jobs in the South, and there is no doubt that the incomes of the residents have improved, but the growth is not at a rate that enables these countries to catch up with those in the North in terms of income. In addition, the motivation to relocate is not to industrialize the South but rather to move manufacturing firms closer to the source of primary resources, cheap labor, and large markets.

8.3.2 Modernization Theory and the North–South Economic Gap

Modernization is an American and Western European concept that was presented as a vehicle for less-developed countries to leave underdeveloped conditions and achieve economic growth (Berger 1994). Policies adopting various frameworks for state-to-state economic development to implement shared growth and improve efficiency and equity have been proposed. Some of the key policies have concerned financing infrastructure development through loans and capital flows to the South. These policies have been implemented by various organizations to address functional problems (Mosley 2005). The challenge is that these policies are implemented and steered by private sector banking institutions that are outside the state-to-state or government-to-government frameworks and whose main motivation is not to achieve shared growth and therefore reduce global inequality but rather to accumulate wealth.

Most people believe that the foundation for economic development in any country is quality, efficiency, and a good transport network and communication infrastructure. Domestic tax revenues often fall short of the expenditures required to develop public infrastructure in countries in the North as well as in the developing countries in the South. For example, in many countries, the national infrastructure costs exceed the national revenue: in Nigeria by a factor of 12, in Ethiopia by 20, and in the Democratic Republic of Congo by almost 26 (Fuss et al. 2016; World Bank, 2014). A study carried out by Kodongo and Ojah (2016) found that both spending and access to infrastructure influence economic growth and that the quality and quantity of support for economic growth are influenced by export diversification and cross-border capital flows. This study does not go far enough to compare how much exports from the South benefit the Southern countries in the form of returns and how much they benefit the importing countries in the North. Such a comparison is necessary, however, because the money used for developing infrastructure in the South is not free, and yet the said infrastructure is designed to provide companies from the North access to cheap labor and market share advantages.

The North–South economic gap is also a result of the fact that state-to-state shared economic rules are heavily influenced by developed countries and their private sector institutions. These rules, according to Mosley (2005), "tend to privilege efficiency over equity and provide more benefits to wealthy nations." Other studies argue that even when countries in the South are influenced by domestic interests and strategic calculations, these may reflect financial sector priorities, which are again dominated by financial institutions from the North (Mosley 2005). The moral principle developed by Singer (1972) suggests that when the reasons for giving aid to poor countries go beyond providing assistance, it should be a moral duty to provide such aid. Classifying international assistance to poor countries as a moral duty means that nothing else, including interest on loans, should outweigh the aim to assist the poor countries (Kaya and Keba 2011), especially when the lenders also benefit through access to much-needed natural resources from the South.

The most economically powerful states have preferred an ad hoc arrangement on FDIs (Mosley 2005) and refused to follow formal, legal relations that support the needs of all parties (countries in the North and South). Many countries in the North have continued to relocate their manufacturing enterprises to the South mainly for purposes of using cheap labor and moving closer to the markets. I am not sure whether to describe this as smart or greedy, but the developed countries of the North have kept their manufacturing advantage by providing a false impression that they are transferring capital to developing countries (Revenga and Dooley 2019). The ownership of all Western-owned manufacturing enterprises remains foreign and serves the interests of shareholders and the home country.

8.4 Methods

There is no question that in the free market and globalized economy, the material flow of resources has been more from the South to the North. This is because high levels of per capita income are accompanied by high levels of resource consumption (Schaffartzik et al. 2019), and by that standard, the North is consuming more material resources than it has. In contrast, the resource use of poor countries is so low that the most basic needs, such as food and shelter, often go unfulfilled (Schaffartzik et al. 2019; World Bank 2018). A study by the Ecological Footprint Network (2016) measuring the natural resources a given country requires to meet its domestic consumption has consistently found that all high per capita income countries have large natural resource deficits, while low-income countries have surpluses. It is therefore fair to argue that because of the global resource-use imbalance, high-income countries that consume higher levels of global resources should compensate low-income countries for making these resources available.

Table 8.1 shows countries from two regional economic blocks (the European Union—EU and the African Union—AU) that have a long history of trade and shared economic development interests. The top five and lowest five countries in terms of per capita income from each bloc are indicated to show the income inequality and natural resource inequality between individual countries. In detail, Luxembourg has the highest income per year in the EU, with US$ 116,640 per capita, and Botswana has the highest income in the AU, with US$8,259 per capita, a difference of US$ 108,381. On the other hand, the Netherlands has the lowest income in the top five from the EU, with US$ 53,024 per capita per year, and Burundi has the lowest income in the selection from the AU, with a mere US$ 272 per year, a difference of US$ 52,752. Biocapacity measures nature's capacity to meet the human demand for natural resources. Countries whose consumption of natural resources exceeds the capacity of nature to meet the human demand are defined as having an ecological deficit (Global Footprint Network 2016). The FDI, grants, loans, and other financial flows within and across countries in the EU and the AU are analyzed.

The biocapacity column shows how much the demands of an individual country are above or below the capacity of nature to meet those demands. For example, if everyone in the world lived like citizens of Luxembourg, we would all need approximately 7.922 Earths per year to meet the human

Table 8.1 GDP per capita income (US$) and biocapacity (Earths) comparison

EU Countries			AU Countries		
Country	**Capita Income**	**Biocapacity**	**Country**	**Capita Income**	**Biocapacity**
High			**Low**		
Luxembourg	116,639.9	7.922	Burundi	271.8	0.404
Ireland	78,806.4	3.140	Somali	314.5	0.595
Denmark	61,350.3	4.175	Malawi	389.4	0.453
Sweden	54,6608.0	3.962	Niger	414.0	1.019
Netherland	53,024.1	2.965	Central African Republic	475.7	0.747
Low			**High**		
Romania	12,301.2	1.898	Botswana	8,258.6	1.670
Croatia	14,909.7	2.415	Gabon	7,952.5	1.420
Poland	15,420.9	2.717	Libya	7,241.7	2.290
Latvia	17,860.6	3.901	South	6,374.0	1.934
Lithuania	19,153.4	3.416	Africa	5931.5	1.626
			Namibia		

Source: World Bank Data, 2018 & Global Footprint Network 2016.

demand for natural resources, a clear deficit of 6.922 Earths. At the other extreme, if everyone lived like citizens from Burundi, we would all need 0.404 Earths per year to meet the human demand for natural resources, leaving a surplus of 0.596 Earths. The amount of demand that citizens put on nature has a strong connection to the country's level of income. High-income countries exert very high demand on natural resources, while low-income countries have very low demand. The imbalance in natural resource use has not been factored into global economic inequality. Given the way the North values its technology and manufacturing, if it had a surplus of natural resources, it is possible that the South would be made to pay for them.

8.5 Findings

As shown in Table 8.1, many countries lack sufficient natural resources within their borders to meet their domestic needs. There is a need to work out a formula of sharing natural resources that takes into account the real contribution of those resources to a specific economy and estimate the opportunity costs of having no access to those resources. The estimate of opportunity costs could be then applied to various development assistance loans given to countries in the South. The sharing of resources should not be the only justification. Kaya and Keba (2011) and Wilkinson and Pickett (2014) argue that inequality should be rejected not because equality is more desirable, but rather because global inequality subverts the fair functioning of public/private international institutions and governments, thus reducing the ability to see the need to integrate national poverty into global inequality.

8.6 Debt and Inequality—What Does History Show?

The flow of FDI, loans, and other grants between the EU and AU countries in Table 8.1 was analyzed and is shown in Figures 8.1 and 8.2. Countries with the lowest GDP per capita incomes also have the highest debt-to-GDP ratio, and their access to external borrowing is dictated by global economic performance. When the global economy is not doing well, countries in the South either stop borrowing or simply cannot be lent loans and Foreign Direct Investments (FDI) flow stops. Global economic performance does not affect countries in the North; in fact, these countries obtain access to more loans when the global economy is not healthy. UN (2009) delegates to the second committee on economic and finance presented information on the impact of the economic crisis of 2007–2010 where "financial resources, Foreign Direct Investment (FDI), remittances from overseas workers, export earnings, and tourism revenue all dried up at the same time." This statement is confirmed by IMF data shown in Figures 8.1 and 8.2.

For example, except during the recession period between 2007 and 2014, the African countries with the lowest per capita income all had a

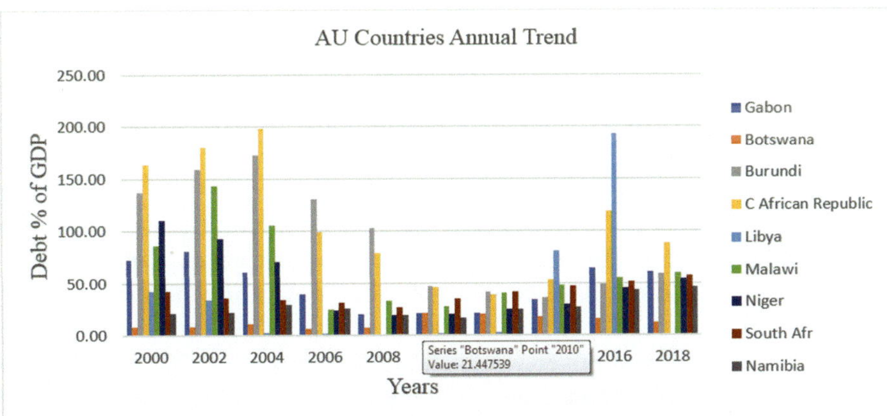

Figure 8.1 Government debt AU.
Source: IMF Data.

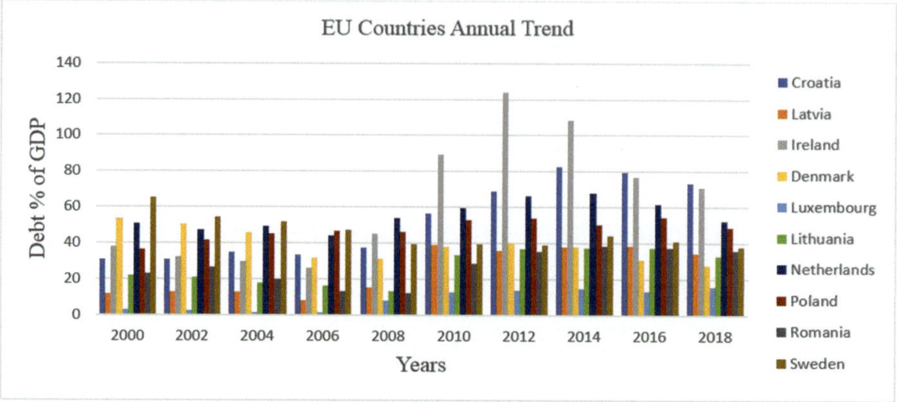

Figure 8.2 Government debt EU.
Source: IMF Data.

debt-to-GDP ratio exceeding 50%. Between 2006 and 2014, the debt-to-GDP ratio of all these countries declined to just below 50%. On the other hand, among all sample countries in the EU, the debt-to-GDP ratio rose from 2006 through 2014 and remained at approximately 40%. When the poorest countries in Africa needed help most, during the recession, they did not receive help, while many countries in the EU saw their debt-to-GDP ratio increase. The period between 2006 and 2014 shows how the debt-to-GDP ratio declined for countries in the South, while the ratio rose and remained slightly higher than before the recession for most countries in the North. "The global stimulus package was felt most in the developed countries" (UN 2009).

In the AU, during the recession period from 2007 to 2014, Gabon, Burundi, Central African Republic, Malawi and Niger, the debt-to-GDP ratios dipped to below 50%. These are the countries with the lowest per capita income, and before the recession, their debt was over 100%. While a high debt-to-GDP ratio is not economically healthy, it is morally bad if the rate drops by hurting people through reduced access to basic needs. Countries borrow because they do not have enough savings from various income sources to finance public expenditure needs. While excessive debt can be an obstacle to economic growth and poverty reduction (Siddique et al. 2015) and while many studies have been concerned with the impact of debt on the economic growth of borrowing countries (Jayaraman and Lau 2009; Clements et al. 2003), this study addresses debt issues from a moral standpoint.

8.7 Approaches to Managing Inequality and Debt

There have been four main paths pursued so far to reduce North–South inequality: (i) the transfer of wealth from more-developed to less-developed countries, (ii) political independence, (iii) the Washington Consensus, and (iv) the 2010 Seoul Development Consensus on Shared Growth by the G20 countries (Zondi 2013). These four approaches are grounded on the concept of global governance and use finance and technology as the vehicle to deliver development outcomes. The participation of the private sector in the global North in facilitating the way finance and capital flows to the global South is not organized on the principles of equality and fairness but rather on efficiency and "what is in it" for those private sector corporations. Therefore, these efforts continue to present mixed results regarding inequality.

These four approaches fail to represent tangible initiatives to reduce global inequality. The conceptions of economic development based on the location of industry as one way to transfer wealth to eliminate global inequality have been inadequate and costly to countries in the South. Attention to location without regard to ownership has been found to lead to a disproportionate percentage of income going to owners, who are mostly foreigners, and not to the local economy (Fowowe and Abidoye 2012). Such an outcome does not translate into industrialization and reduced inequality between individuals and between countries. Therefore, the conventional approach to reducing global inequality needs to be organized around real estimates of the unbalanced opportunity costs between financial surpluses/deficits and natural resource surpluses/deficits. This means developing relationships with foreign industries that exert positive effects on economic growth, such that the profits made by those foreign firms as a result of their locational "local" advantages are able to benefit the local economies as much the shareholders in the home countries.

The Washington Consensus, which was presented as a platform for inclusive, shared, and sustainable economic growth and development, has instead, as described by Zondi (2013), through capitalism turned the South into a source of cheap labor and capital for the North. Under the Washington Consensus, the South was turned into a "*structural periphery in a growing and affluent global North through the dispossession of land and natural resources, the expropriation of profits and the exploitation of labor*" (Zondi 2013). Unfortunately, many countries that fought for independence and obtained it have also been blind to this reality, as shown by the extent of the North's control over these countries (Zondi 2013).

The free market model with unfair competition continues to be presented as the best vehicle for economic development. This can only be true in the private sector, where the sole mission is profit making. The challenge is that the private institutions in the global North and the governments in the South do not have the same capacity, resources and ingenuity to ensure a shared growth, such that the profit interests of the private sector overshadow all other intentions. Atisa et al. (2020) found that in places where there are conflicts regarding the use of natural resources for different purposes by different stakeholders, unequal competition emerges, and sometimes, dispossession results, carried out by international corporations with the money and power to manipulate the decision-making sphere away from equity issues. Most studies argue that global inequality has been declining in the last 20 years (Revenga and Dooley 2019), but this is mainly because the industrial gap in the global South is being filled by the global North.

Inequality cannot decline in a global economy where wealth and natural resource benefits continue to accrue to a "few stakeholders," leading to a deteriorating poverty and wealth distribution (Atisa et al. 2020). This has much to do with the ownership of capital. Many of the industrial firms in the South are owned by firms from the North that eventually repatriate much of the income and profit back to their home countries. Many of the foreign firms operating in developing economies are owned by private entrepreneurs from the North and retain the design, innovation and critical knowledge functions in the North. A study performed in the Philippines by Bonito et al. (2017) found that entrepreneurship had no significant impact on poverty reduction and income inequality. On the other hand, the same study found that GDP was a significant factor in determining income inequality, meaning that if firms are not contributing significantly to GDP, inequality cannot be expected to decline.

8.8 The Costs of Modernization for Countries of the South

Lee et al. (2013 pp. 524) argue that a debt service to expenditure ratio of 5% is good, and approximately 10% is acceptable for the state governments in the United States of America. The growing debt of countries in the South, among other variables, contributes to North–South inequality. In countries whose domestic revenue margins are so thin, unlike those in the states in the USA, a high debt-to-GDP ratio leads to the diversion of scarce financial resources from funding domestic programs to servicing loan obligations (IMF 2019). In an IMF (2019) study of the debt-to-GDP ratio of low-income countries, an increasing interest rate burden as a result of the increasing volume of loans was found.

The labor-saving innovations continue to impact the labor demand for skilled and unskilled workers and, hence, the incomes in the South. Savings made from automation mainly go to the top skilled labor and owners of the enterprises, thus widening the income gap (IMF 2019). First, the evolving governance framework serves the narrow interests of investors coming from the North and not the states in the South that host the relocated industries. The voice of countries in the South is limited to the extent allowed by the private sector institutions from the North, leading to relations that are founded not on the allocation of value to address inequalities but rather on profits for foreign enterprises. Many countries have continued to experience economic growth, but the same growth has not translated into human progress and national prosperity (Zondi, 2013). Again, this is the reason why there is a narrowing gap in industrialization that does not translate into a narrowing gap in incomcs.

The public needs notwithstanding, corruption and institutional weaknesses drive developing countries to seek more loans without much regard to their capacity to manage debt. Debt capacity refers to how countries are able to manage their expenditures and resources and governments' ability and commitment to meet the debt requirements (Lee et al. 2013, pp 523). *"High debts and deficits, along with the associated financing requirements, leave countries vulnerable to interest rate and fiscal risks and may be a drag on long-term growth"* (IMF 2019).

8.9 Discussion

8.9.1 What Can the South Do?

Countries in the South know very well that they have not obtained the desired economic outcomes from the transfer of wealth, relocated industries and loans given by developed countries from the North and other international lenders. One of the challenges they face is making the North pay for the use of their surplus natural resources in exchange for development assistance of various forms. In the face of such difficulties, a pragmatic and morality-based framework remains the only viable and acceptable approach to negotiating shared growth agreements. International institutions may be able to help by developing a scale for evaluating North–South wealth and resources exchange mechanisms.

In a liberalized global market, participants do not take turns seizing opportunities. Entrepreneurs have been found to emerge from groups of people who feel they are deprived of opportunities in terms of access to financial resources or some type of welfare (Jenkins 1983). People with very few resources and minimal political and organizational experience have been found to create durable and impactful innovations. Entrepreneurs use what is known as episodic knowledge to generate value when markets present opportunities; they do not require ownership of resources but rather ideas that add context to uncertainty (Atisa 2020; Simpeh 2011; Knight, 1921). Clearly, this is what the North has been doing, taking advantage of natural resources from the South. Through an entrepreneurial approach, the South must find ways to take advantage of the financial resources of the North. Because both regions have entrepreneurial capacity and understand what they want from each other, the North–South economies will be forced to develop a shared growth model that will benefit them both and eventually lead to reduced inequality.

There has also been a push for free trade and open economies mainly from the North. In more open economies, there are fewer regulations constraining participation in the global business environment, which has become a winner-takes-all community. Looking at the extent to which competition has intensified, economic openness can be described as one of the avenues leading to global inequality. Simpeh (2011) explains opportunity-based theory, according to which entrepreneurs do not cause change but are able to exploit opportunities created by change itself. "Entrepreneurs have an eye for the possibilities created by change, not the problems" and are constantly looking for and responding to changes by exploiting them as opportunities. Jenkins (1993) asserts that entrepreneurship easily emerges among deprived groups and factionalized communities. These are the type of conditions currently existing in most communities in the South.

If the observation by Jenkins is correct, countries in the South must start to ask themselves why they are unable to develop the entrepreneurial skills necessary to start generating wealth and replace the need for imported development capital from the North. This could be done through developing economic policies that mitigate specific aspects of development assistance from the North that do not serve their interests. Moreover, efforts tackling economic problems in the South should move beyond the theoretical discussions about inequality to begin addressing the practical world of global competition, the firm grip of capitalism on the global economy and the global society of winner-take-all corporate policies that are here to stay. The only solution is a global society guided by a moral compass to address inequality as a moral problem and not as a business outcome or competition problem.

8.9.2 What Can the North Do?

One question the North should be asking is, given the current high deficits in biocapacity and natural resources, what would be the cost of obtaining the same resources from alternative sources (there are no known alternative sources)? Some advanced countries in the North are currently investing in technology to go as far as the planet Mars in a search for human habitable conditions. If there were habitable conditions, they could not be shared with planet Earth unless everyone from the North moved to Mars. There is nothing wrong in investing to expand the frontier of living conditions, but the reasons behind such investments and how the investments could serve the entire population of the region are worth examining. The question such efforts ignore is, what if these efforts and funds were invested in sustainable development initiatives in countries with surpluses in biocapacity to make sure there is a continued flow of resources to wherever they are needed? Answering this question would require prioritizing expenses and benefits so that countries can start to invest in a more equitable society and not in wealth, power, and privileged development frameworks.

The free market economic framework that currently governs economic relations globally is not about fairness but is rather a justification for the rich, most powerful and most innovative to maximize the returns from the businesses they own against all other competitors. It is called fair competition, but from a philosophical lens, this conceptualization of fairness under the free market is responsible for global inequality. The free market framework does not care about losers; it is purely an expected outcome that there must be losers and winners. How the winners in a free market economy treat losers is something that has to be addressed by the market participants and not the market itself. This issue requires a cosmopolitan approach, especially for industrial firms that have relocated their operations to the South, to finding a balance between profit margins and the investment in local economies. To be a cosmopolitan is an inspirational ideal which, according to Nath (2011), "is to see oneself as a part of a global community comprising all of humanity, identifying as a member of one's local community."

This means that the private sector or the markets should not be allowed to dictate transnational help based on commerce but rather that assistance should be government led and based on community needs. Using the philosophy of John Rawls regarding global economic justice, Miller (2011) explains the conception of fair equality of opportunity dictated by political duties in support of "the lifetime expectations of income and wealth of the worst-off" participants in a project. The "worst-off" participants in this case are the businesses from the South and the governments that agree to host multinational corporations from the North with a hope of advancing toward industrialization. Legally, corporations are regarded as people, and therefore, they should have a feeling toward the communities in which they find themselves through regulations. Global corporations should identify with the local communities where they have relocated, accept being subjected to political duties in the host country and understand the material expectations and needs of the country.

8.9.3 Who Else Can Take Responsibility?

Inequality is not a problem to be left to governments and private sector initiatives. It needs to be defined broadly, and more players with an interest in serving the greater public good invited to take part in developing relations between governments and private sectors in the North and the South. The existing relationships leave the South at the mercy of powerful Northern interests. The decisions made in the North are more binding in the South, while decisions made in the South have no influence on the way Northern governments run their economies and business. There must be a third player with no vested interests on either side.

With regard to making regulatory rules on global financial flows and investments, the private sector has had greater influence than governments or public sector institutions over how decisions are made (Mosley 2005) in both the North and the South. Unfortunately, it is mainly the global corporations from the North that continue to present such great influence, with almost zero input from the industrial sector in the South. It is right, therefore, to assume that with the insignificant influence of the South, human values and the understanding of how economic arrangements affect communities in the South are not well understood globally.

Global relations should go beyond politics and economics to include human values with rising global connections and interdependence. This is because regardless of the direction of capital flows, those that invest more tend to gain more, and these actors should be concerned about giving back to the local communities where their investments are generating the greatest returns. Keohane and Nye (1971) raised important questions on the allocation of economic value, who benefits from international relations and how this impacts inequality between countries. Mosley (2005) hypothesized that private sector "institutions tend to serve the interests of powerful countries and powerful constituencies within those countries, thus limiting progress in equity." For these reasons, countries from the South have no viable platform to voice their displeasure with the biased, predetermined decisions except through other international players.

I propose that international institutions, such as the UN, World Bank, IMF, and World Trade Organization, restructure their global policies and take a stronger role in the search for ways to reduce global inequality. For a long time, these international institutions have paid attention to the global distribution of wealth and power (Mosley 2005), but they have done so through the lens of the Northern private sector economic policy of free market capitalism. In a more globalized world, these institutions have continued to expand their role beyond economic and political relations into the realms of human rights and global security. Because the concept of human rights and global security has been narrowly defined, global inequality has not been perceived as a human rights and global security landmine. Many human rights abuses and global security problems have their origins in poverty, perceived unfair treatment by the wealthy and powerful, and global corporate greed. International institutions can facilitate the fair treatment of all countries on a global economic platform through the creation of rules where a percentage of profits from global investments can be channeled toward reducing global inequality.

8.9.4 Reconciling North–South Development

The North–South inequality is not an abstract issue but a real challenge that is felt by communities in the South through real deprivation, dispossession, and unfairness but seen by the North as a market opportunity to be exploited. Global inequality cannot be reduced by investing in the expansion of own industries and opening up branches in poor countries but rather by investing in reducing barriers to growth in the South. There are good and moral reasons to be troubled by the large economic inequalities between countries and between individuals globally. The character of human interaction and the ways of doing business should not solely be guided by the invisible hand of the market. Nath (2011) argues that "well-off citizens of the world have obligations to reduce global inequality," and they can meet these obligations through adopting fair business practices.

The starting point would be reconciling policies that are designed to address corporate-centered needs and policies that address human-centered needs in this North–South inequality. Governments from the South are clearly struggling to make profit-centered private sector stakeholders from the North see the human-centered needs and support development in the South. The North and South lack a common focus on what is and should be a shared responsibility outside of their individual missions or core principles. There are no tools ready for the South to deploy in order to create a common focus among all stakeholders in this North–South relationship. This means examining the past and potential future relationship and exploring how to achieve a more equal relationship. The aim is to build a global society where the North and South can realistically influence each other, such that the North alone cannot shape the South, and the South alone cannot shape the North, different from the current framework of inequality.

The North has adopted FDI as the means for transferring capital to developing countries. This is merely a strategic effort designed to expand the market and lock in market share (Moran 2011, page 34). Rather than opening full-fledged production firms in the South, corporations from the North have opted for assembly lines to save on the shipping costs of completely assembled machinery. Moreover, in addition to providing access to cheap labor, this means allows corporations to set up plants behind trade barriers and become more competitive in expanded markets (Moran 2011). It is possible, although unproven, that the incomes paid to local laborers in these assembly firms represent a small percentage of the savings made relative to the case if these companies were to ship completely assembled machinery. By setting up assembly lines where local labor is only involved in putting parts together, there is no real technical knowledge transfer, as there would be if these companies actually built production centers. This approach can be expected to block any path toward industrialization and should be abandoned.

Another way to try to reconcile North–South development is to take natural resource use beyond environmental implications and try to determine the economic contribution toward specific global, regional, and individual economies. Until global economies establish a clear linkage between material use and economic growth, reconciling North–South development will be difficult. Countries of the South are endowed with rich natural resources, while countries in the North boast high levels of technological development. In addition to having rich technological endowments, countries in the North are more advanced in economic terms than those in the South. The mutual importance of technology and environmental resources to each other has not been recognized. There needs to be a way to determine in terms of value how much natural resources from the South contribute to the technological advancement in the North and share the benefits of this contribution.

8.10 Lessons Learned

Kakwani and Silber (2008), writing about human development in a special issue devoted to multidimensional poverty analysis, quoted Arthur Lewis (HDR, 1996, p. 46) and stated that "development means widening the range of human choices." Looking at poverty from a development perspective, I like the conceptualization provided by Kakwani and Silber (2008) and would like to explore some lessons from this view to enrich the thinking on inequality.

On the one hand, poverty can be defined "in absolute terms, where it refers to income levels that do not guarantee that basic physical needs are covered." Poverty can also be defined in relative terms, as income levels that do not allow individuals to function properly in their social environment. Global inequality is defined broadly as the disparity in incomes. However, income disparity does not tell the full story; instead, inequality concerns the levels of income that cover basic needs and allow individuals to function properly in their social environment. These definitions suggest that people should be given opportunities under the shared growth framework to make their own incomes that cover basic needs and enable them to function properly in their social settings. In an open market and a competitive world, this speaks to specific efforts and types of help that should be provided by rich countries from the North to enable the South to function properly.

More studies focus on how multinational corporations can tap into their full potential and gain maximum benefits in foreign markets (Ferdows 1997) than on how host countries can also benefit from foreign firms. Studies have found that before the local markets in the South started burgeoning, FDI was mainly intended for re-export back to the North (Johannes Van Biesebroeck and Timothy Sturgeon 2010). High-value functions, such as innovation and design (knowledge-intensive), remained in the North, while production activities migrated to the South. This, however, is not the case across countries in the North, where manufacturers do not place much importance on wages and market share but willingly and easily allow FDI flow, knowledge, and productivity among each other. Ferdows (1997) found that when manufacturers do not move technology with complete design and knowledge-based functions, the host locations are provided with limited resources and a narrow strategic scope that are not beneficial. This is the reason we see many manufacturing firms from the North establishing assembly lines only in the South. In such a framework of industrialization that lacks the transfer of innovation, design and knowledge processes, economic inequality cannot be solved.

The model using development loans and grants to develop infrastructure in the South cannot guarantee the reduction of inequality. This is because public infrastructure spending can have positive effects on economic growth only under very specific circumstances, when there is strong complementary spending from private capital, in the South (Easterly and Serven 2003). The North therefore needs to invest in private sector partnerships in the South and create a sector that supports local infrastructure development needs. The grants, handouts or emergency assistance that the North has been providing will not change the status quo until there are ways to start investments that target the development of individual and productive sector capacity and reduce the vulnerability to expectations of handouts or loans. Pro-poor growth policies that can rid the world of inequality cannot come from handouts, loans or relocating manufacturing in the absence of clear and tangible efforts to address the underlying capacity challenges in the South.

There are many potential and untested ways to try to eradicate global economic inequality. First, national governments in the North have given power to the private sector to make business sector rules, and these businesses seem to exercise this power globally. Governments from the South seem too willing to be bound by private sector rules originating in the North. The current market-oriented framework cannot be trusted to reduce global inequality. On a planet where humanity is clearly facing shared problems, such as the ongoing COVID-19 pandemic, climate change and terrorism-related threats, the call to reduce global inequality has never been more urgent. The steps each country and region take to address these shared problems can add to or subtract from the global outcomes and the global effort. Every country on earth should have the capacity to face these global threats. This is only possible in a global order where political institutions are able to mandate economic institutions to balance profits with service to the poor (Nath 2011).

Development needs of countries from the South were not prioritized during the financial crisis of 2007–2010. The global economy, in the year 2020 is again going through another financial crisis brought about by COVID-19, an infectious disease that has spread to the entire globe. The global economy has been shut down in order to stop the spread of this infectious disease. There is no doubt, if the approach to recovery once this crisis comes to an end follows the same framework like the 2007–2010 financial crisis, the countries from the South are likely to sink deeper into poverty, more than they did in the period after the last global crisis. The recovery was much slower and the FDI to South declined by big margins as shown in Figures 8.1 and 8.2. This current crisis is hurting all countries more than the previous one and how long the global recovery will take is difficult to predict. It is likely to leave the South more isolated economically than before. Based on the moral principles and the current incomes/resources, deficits/surpluses, the AU, together with the international development institutions should work out a new framework to ensure a fast, steady, and broader recovery for countries in the South once this crisis is over.

8.11 Conclusions

North–South inequality has been found to aggravate poverty and is a moral, political, and economic issue in a globalizing world that needs attention

(Kacowicz 2001). The real indicators of narrowing global inequality are the differences in the labor income between individuals in the North and the South and the strength of the linkage between the per capita income in Northern countries and the biocapacity levels in the South. At present, the inequality between individuals in the North and in the South has continuously been rising (Arrighi et al. 2003), and this is not merely a contradiction of the literature but rather the global reality.

The current framework of wealth distribution is guided by power and privilege, making it very difficult to reduce global inequality. Wealth does not come in the form of money alone but can be seen in other forms, such as natural resources; likewise, inequality can also be classified into various forms, and all those forms need to be considered on the same scale. There is no justification for global inequality when every region has something to offer that makes the entire world a better place. It should be a moral aspiration of the North to work toward equity as long as such efforts do not pose economic costs. Globalization, if left unchecked, can be detrimental to the global distribution of resources, bringing about greater inequality rather than reducing it. This is clearly displayed by the differences in per capita income and biocapacity between the countries in the North and South. It is right to conclude that as long as the private sector from the North continues to have a free hand in the rule-making regarding international finance, global inequality will continue to increase rather than shrink.

Investments by the North (high per capita countries) in the South (low per capita countries), which have long been presented as enabling development, are in reality an extension of the resource consumption possibilities frontier that has little to do with development in the South (Schaffartzik 2019). High-income countries lack sufficient natural resources within their national borders, as has been estimated by the Global Footprint Network (2016), and have to rely on the resources in the South. This provides a moral and pragmatic justification for the North to assist the South in getting out of poverty.

There are many legitimate international organizations designed to influence global affairs for a just society, such as the UN, the IMF, the World Bank, and the WTO. The small extent to which these have been able to influence the powerful and rich countries and their private sector is worrisome. These international institutions, which should provide solutions to functional problems regarding distributional and economic benefits between countries, have failed to take a middle-ground strategy to guarantee equitable relationships. For example, as trade and financial interdependence have expanded between the North and the South, the preferences of the North have also gained greater influence over decisions that affect both regions (Mosely 2005). Historically, rather than adjudicating fair international relationships, the existing international institutions have provided a way for the rich countries to impose their preferences regarding global economic challenges (Mosley 2005; Drezner 2003; Simmons 2001), and this must tangibly change.

References

1. Arrighi G., Silver B. J., and Brewer B. D. (2003). Industrial Convergence, Globalization, and the Persistence of the North-South Divide, *Studies in Comparative International Development, Spring 2003, Vol. 38, No. 1, pp. 3–31.*
2. Atisa, G., Aziza, A and Weiss M. (2020). Decentralized Governments: Local Empowerment and Sustainable Development Challenges in Africa. *Development, Environment and Sustainability.*
3. Atisa, G. (2020). Policy adoption, legislative developments, and implementation: the resulting global differences among countries in the management of biological resources. *International Environmental Agreements* 20**,** 141–159.
4. Atisa, G. (2020). Resource Curse in Kenya's Coastal Region: A Symptom of Institutional Failure? The end of location theory *Towards Better Framework for Managing Africa's Coastal-Hinterland Divide, Blue Economy & Maritime Disputes (forthcoming).*
5. Beitz C. R. (2005). Cosmopolitanism and global justice. *Journal of Ethics*, 9: 11–27.
6. Berger M. T. (1994) The end of the 'third world'? Internationalisation and the History of 'Development'
7. Bhandari, Medani P. (2020) Second Edition- Green Web-II: Standards and Perspectives from the IUCN, Policy Development in Environment Conservation Domain with reference to India, Pakistan, Nepal, and Bangladesh, River Publishers, Denmark / the Netherlands. ISBN: 9788770221924 e-ISBN: 9788770221917
8. Bhandari, Medani P. (2020), Getting the Climate Science Facts Right: The Role of the IPCC (Forthcoming), River Publishers, Denmark / the Netherlands- ISBN: 9788770221863 e-ISBN: 9788770221856
9. Bhandari, Medani P. and Shvindina Hanna (2019) Reducing Inequalities Towards Sustainable Development Goals: Multilevel Approach, River Publishers, Denmark / the Netherlands- ISBN: Print: 978-87-7022-126-9 E-book: 978-87-7022-125-2

10. Biesebroeck J.V. and Sturgeon T. (2010). The automotive industry in developing countries. CERP Policy Portal. https://voxeu.org.
11. Bonito, J.D.M., Daantos F.J.A., Mateo J.C.A. and Rosete M.A.L. (2017). Do Entrepreneurship and Economic Growth Affect Poverty, Income Inequality and Economic Development? Review of Integrative Business and Economics Research, Vol. 6, no. 1, pp. 33–43.
12. Clements, B., Bhattacharya, R., and Nguyen, T. (2003). "External Debt, Public Investment, and Growth in Low-Income Countries", IMF Working Paper No. 249, WP/03/249, IMF, Washington.
13. de Soyres C, Rogantini P. A., and Sab R. (2019) Debt Build-up in Frontier Low-Income Developing Countries (LIDCs) since 2012: Global or Country-specific Factors and Way Forward? IMF Working Paper Finance Department.
14. Drezner D. (2003). "Who Rules? The Regulation of Globalization." Manuscript, University of Chicago.
15. Duro J. A. and Teixidó-Figueras J. (2013). Ecological footprint inequality across countries: The role of environment intensity, income and interaction effects. *Ecological Economic 93: 34–41*
16. Easterly W. and Servén L. (2003). The Limits of Stabilization: Infrastructure, public deficits, and growth in Latin America. The International Bank for Reconstruction and Development / The World Bank
17. Ecological Footprint Network (2016). National Footprint Accounts 2016 are out! Carbon makes up 60% of world's Ecological Footprint. https://www.footprintnetwork.org/2016/03/08/national-footprint-accounts-2016-carbon-makes-60-worlds-ecological-footprint/
18. Ferdows K. (1997). Making the Most of Foreign Factories. Harvard Business Review
19. Forowowe B. and Abidoye B. (2012). The effect of financial development on poverty and inequality in african countries. *The Manchester School* Vol 81 No. 4 562–585 doi: 10.1111/j.1467-9957.2012.02302.
20. Fuss S., Chen C., Jakob M., Marxen A., Rao N. D., and Edenhofer O (2016). Could resource rents finance universal access to infrastructure? A first exploration of needs and rents. *Environment and Development Economics*, Available on CJO 2016 doi:10.1017/S1355770X16000139
21. IMF (2019). Fiscal policy for a changing global economy.
22. Jayaraman, T.K., and Lau, E. (2009). "Does External Debt Lead to Economic Growth in Pacific Island Countries", Journal of Policy Modelling 31: 272–288.
23. Jennkis J. C. (1983). Resource mobilization theory and the study of social movements. Annual Review of Sociology, 9:527–53

24. Kakwani N. and Silber J. (2008). Introduction: Multidimensional Poverty Analysis: Conceptual Issues, Empirical Illustrations and Policy Implications, World Development Vol. 36, No. 6, pp. 987–991, 2008
25. Kaya A. and Keba A. (2011). Why global inequality matters: Derivative global egalitarianism. *Journal of International Political Theory*, 7(2) 2011, 140–164
26. Keohane R. O. and Nye J. S. (1971). Transnational Relations and World Politics. Harvard University Press.
27. Knight F. H. (1921) 1921. *Risk, Uncertainty, and Profit.* Library of Economics and Liberty. Retrieved fromwww.econlib.org/library/Knight/knRUP1.html.
28. Kondongo O. and Ojah K. (2016). Does infrastructure really explain economic growth in Sub-Saharan Africa? *Review of Development Finance 6. 105–125.*
29. Lee R. D., Johnson R. W. and Joyce P G. (2013). Public Budgeting Systems. Jones & Bartlett Learning LLC.
30. Miller R. W. (2011). Rawls and Global Justice: A Dispute over a legacy. The Monist, Vol. 94 (4), pp. 466–488.
31. Mosley L. (2005). Private governance for the public good? Exploring private sector participation in global financial regulation. Paper prepared for a festschrift in honor of robert o. Keohane conference at princeton university
32. Moran T. H. (2011). FDI in Manufacturing and Assembly: Book Chapter in "Foreign Direct Investment and Development: Launching a Second Generation of Policy Research. Peterson Institute for International Economics (PIIE).
33. Nath R. (2011). Equal standing in the global economy. The Monist, Vol. 94, no. 4, pp. 593 – 614.
34. Pattillo, C., Poirson, H., and Ricci, L. (2002). "External Debt and Growth", IMF Working Paper No. 2, IMF, Washington.
35. Rekha Nath (2011). "Equal Standing in the Global Community. *The Monist,* vol. 94, no. 4, pp. 593-614. THE MONIST, Peru, Illinois 61354
36. Revenga A. and Dooley M. (2019) Is inequality really on the rise? Brookings Institute https://www.brookings.edu/blog/future-development/2019/05/28/is-inequality-really-on-the-rise/ accessed on April 14, 2020.
37. Schaffartzika A., Duroc J. A., and Krausmann F. (2019). Global appropriation of resources causes high international material inequality – Growth is not the solution. *Ecological Economics, 163. Pp. 9–19.*

38. Siddique A and E Selvanathan E. A. (2015) The impact of external debt on economic growth: empirical evidence from highly indebted poor countries. Discussion Paper 15.10. University of Western Australia
39. Simmons B. (2001). "The International Politics of Harmonization: The Case of Capital Market Regulation," *International Organization* 55(3): 589–620.
40. Simpeh, K. N. (2011) Entrepreneurship theories and Empirical research: A Summary Review of the Literature. European Journal of Business and Management, Vol 3 (6).
41. Singer P. (1972). Famine, Affluence, and Morality: Philosophy and Public Affairs, Vol. 1, No. 3 (Spring, 1972), pp. 229–243
42. Therien J-P (1999) Beyond the North-South divide: The two tales of world poverty, Third World Quarterly, 20:4, 723–742, DOI: 10.1080/01436599913523
43. UN (20099. United Nations. Developing Countries, Hardest Hit by Global Financial Crisis, Urgently Need Pledged Assistance, Assembly's Economic Committee Is Told, Sixty-fourth General Assembly, Second Committee. https://www.un.org/press/en/2009/gaef3248.doc.htm.
44. Wilkinson R. G. and Pickett K. E. (2014). The World we need. International Journal of Labour Research, Vol. 6 (1).
45. World Bank (2016). *World Development Indicators 2016*. Washington, DC: World Bank.doi:10.1596/978–1-4648–0683–4.
46. World Bank (2018). World Inequality Report. https://wir2018.wid.world/.
47. Zondi S. (2013) The G20 development consensus: an African perspective, Africa Review, 5:1, 73-86, DOI: 10.1080/09744053.2013.832068

9

The Relationship Between Inequality and Socio-Economic Drivers as a Tool for Sustainable Territorial Development

Yuliia Matvieieva, Yuliia Opanasiuk, Olena Pavlenko and Iuliia Myroshnychenko

Department of Management, Sumy State University, Ukraine

9.1 Introduction

The development of sustainability performance indicators and the impact of economic, environmental, and social drivers on the local and global growth have received increased attention in the literature. The 2030 Agenda for Sustainable Development which includes a set of 17 Sustainable Development Goals (SDGs) and 169 global targets, pays attention to "integrated and indivisible and balance the three dimensions of sustainable development: the economic, social, and environmental" (UN General Assembly, 2015). The World Commission on Environment and Development (WCED) in the report "Our Common Future" (the Brundtland Report, 1987) define three main pillars of sustainable development: economic growth, environmental protection, and social equality. In this chapter, the relationship between inequality and socio-economic drivers as a tool for sustainable territorial development is investigated.

9.2 Literature Review and Bibliometric Analysis

A number of literature reviews show that researchers use different terms to describe the interaction between economic, ecological, and social systems. For example, Munasinghe (1992, 1994) apply the term as "sustainable development triangle," Basiago (1999), Purvis et al. (2019) use "three pillars of sustainability." Also, many other researchers have adapted the triangle concept" according to their own research contexts. For example, "dimensions," "models," "components," "aspects" or "perspectives," but the initial meaning of three common characteristics remains the same.

The other terminological problem is to define the type of interconnection model for economic, environmental, and social drivers—balanced, optimal, sustainable, or endogenous growth model.

Using the "title, abstract, keywords" field of search in the Scopus database, search result for "optimal growth" or "balanced growth" (limited to "Economics, Econometrics and Finance," "Business, Management and Accounting") was identified. The search result shows the significantly increasing numbers of papers published as well as long historical background (from 1958) of these research topics. Network visualization of most frequent terms used for "optimal growth" or "balanced growth" is presented in Figure 9.1. This figure points out the strong connection between optimal and endogenous growth. Moreover, clusters for economic, social, and environmental issues for defining optimal growth path are clearly seen.

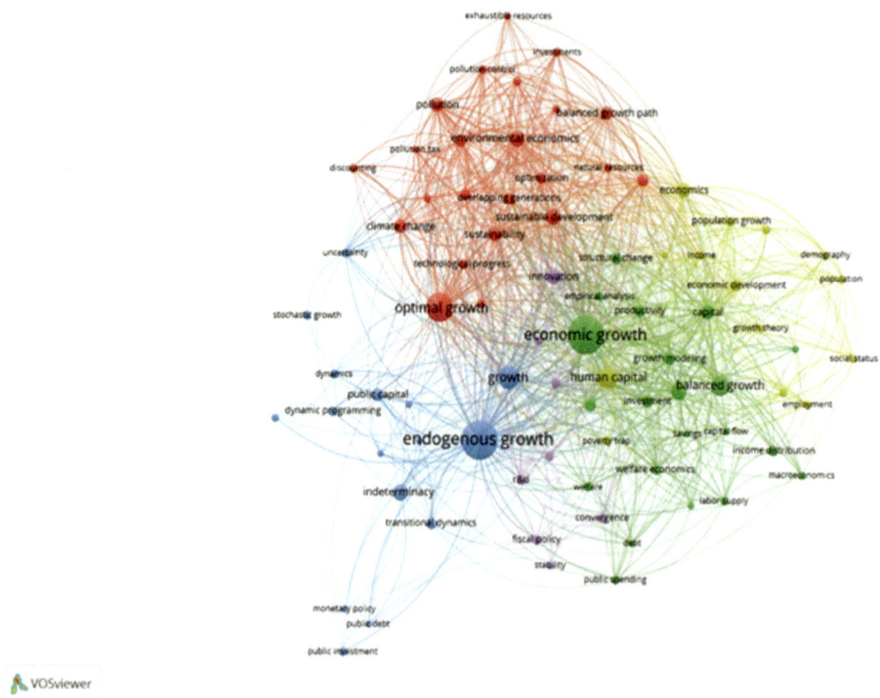

Figure 9.1 Network visualization of most frequent terms.
Source: Authors (Constructed via VOSViewer for the sample N = 1226, f (frequency) ≥ 7).

The United States, France, Japan, United Kingdom, Germany, Italy, Spain, Canada, Australia, and China were among the top ten countries in the world in terms of a total number of publications in the field of "optimal/balanced growth" during 1958–2019. Moreover, it is worth to mention the following top five institutions that work in such research field: CNRS Centre National de la Recherche Scientifique, Université Catholique de Louvain, Kyoto University, Universite Paris 1 Pantheon-Sorbonne, and Cornell University.

Figure 9.2 shows the authors with the highest publication counts (top ten authors). The data confirms that the presented researchers are the ones who have significant research contributions in terms of citation as well.

Greiner (2007) has developed an endogenous growth model with environmental pollution and public capital and demonstrated the existence of a unique and balanced growth path. Stochastic optimal growth with nonconvexities, non-linear dynamics, and chaos in optimal growth are investigated by Nishimura (1983, 2012).

The modern approach of investigating the stimulating factors of economic, social, and environmental development is presented by Barbier and Burgess (2017). Barbier and Burgess (2017). developed a systems approach to sustainability by distribution of three interlinked systems to 17 SDGs. By using methods of the choice theory and welfare, the authors conclude: "that

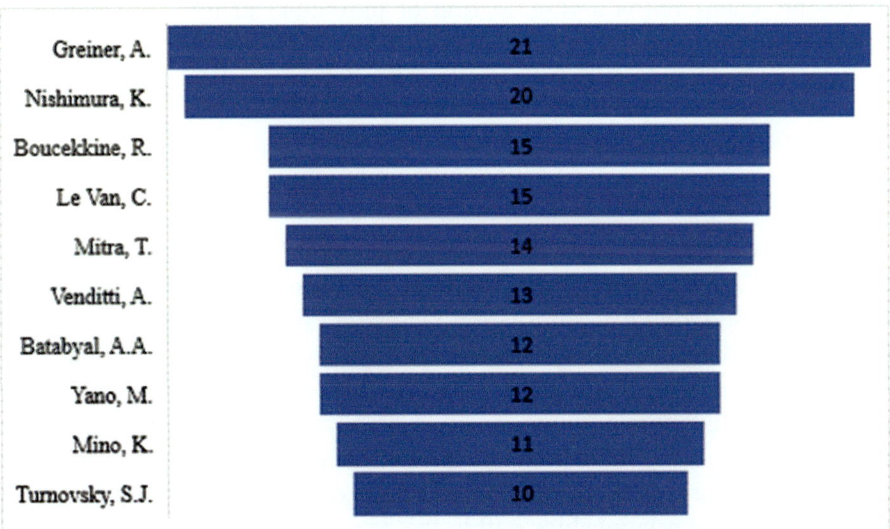

Figure 9.2 Most-published authors during 1958–2019 in the field "optimal/balanced growth."
Source: Authors (Constructed for N = 1226, using SCOPUS database).

it is possible to measure the welfare effects of an increase in the indicator level for one SDG by identifying the tradeoffs that occur while achieving another goal" (Barbier and Burgess, 2017: 9-20). Purvis et al. conclude that sustainability, "through its complex and disparate historical origins, remains both context specific and ontologically open, and thus any rigorous operationalisation requires explicit description of how it is understood" (Purvis et al., 2019: 9-20).

It should be noted that the current scientific literature has mainly addressed the impact of the inequality on the economic and social aspects of sustainable development (i.e., Pavlenko et al., 2020, Telizhenko et al., 2019); whereas, the topic of environmental inequality has not been adequately covered. Our research focuses on the relationship between inequality and environmental sustainability, as well as the relationship between inequality and economic, environmental, and social drivers.

9.3 Data and Methodology

Modern world development trends of countries are aimed at achieving sustainable development goals. This is especially relevant in the context of the unequal development of countries. At the same time, scientists study the various areas of countries' development separately and as a whole. Each country develops in three main areas: economic, social, and environmental. At the same time, each area is a combination of individual factors. These factors include the ones that:

- stimulate the development of the corresponding area;
- destimulate the development of the corresponding area; and
- are neutral.

Special attention is paid precisely to the stimulating factors of the economic, social, and environmental areas of the country's development in the research. These individual stimulating factors are the drivers of the development. It should be noted that drivers are considered as a set of mechanisms that provide the growth impulses, which in turn arise in a particular sector and change the economic, social, or environmental situation of a country in a favorable trend. At the same time, the growth impulses in a particular industry can ensure favorable trends for the development of other industries, taking into account multilevel relationships within the development system of a country.

Thus, the economic, social, and environmental drivers are of particular importance in scientific research ($\sum d^i_{econ}$, $\sum d^i_{soc}$, $\sum d^i_{env}$). They are the characteristics of the corresponding economic, social, and environmental areas of the country's development (N_{econ}, $N_{soc,}$ N_{env}).

The economic drivers ($\sum d^{i}_{econ}$) include the development of entrepreneurship, innovation and investment activities, the development of transport and logistics infrastructure, the development of the agricultural sector, the conditions for doing business, the access to the market and infrastructure, quality, etc. The indicator that most fully describes the state of the economy is GDP. At the same time, GDP is a macroeconomic indicator that reflects the market value of all finished goods and services (i.e., intended for direct consumption or use) produced per year in all sectors of the economy for consumption, export, and accumulation, regardless of national origin of the used production factors. GDP per capita is the most informative indicator taking into account the unequal number of populations in different countries (GDPpc).

$$N_{econ} = f\left(\sum_{i=1}^{4} d^{i}_{econ}\right), \tag{9.1}$$

where d^{1}_{econ} is the Enterprise Conditions, d^{2}_{econ} is the Investment Environment, d^{3}_{econ} is the Market Access & Infrastructure, d^{4}_{econ} is the Economic Quality.

The social drivers ($\sum d^{i}_{soc}$) is a set of indicators that characterize the country's human potential and the living conditions of the population. These include education, skill level, professional skills of the population, ensuring personal freedoms, public administration, living conditions of the population, etc. The indicator that most fully describes the state of the social field is the human development index. At the same time, the human development index (*HDI*) is a comprehensive comparative indicator of life expectancy, literacy, education, and living standards for countries around the world. This index is used to identify differences between developed, developing, and emerging countries, as well as to assess the effect of economic policies on quality of life.

$$N_{soc} = f\left(\sum_{i=1}^{5} d^{i}_{soc}\right), \tag{9.2}$$

where d^{1}_{soc} is the Personal Freedom, d^{2}_{soc} is the Governance, d^{3}_{soc} is the Social Capital, d^{4}_{soc} is the Living Conditions, d^{5}_{soc} is the Education.

Environmental drivers ($\sum d^{i}_{env}$) can include industrial diversification based on cheap renewable energy, the development of the tourism industry, conservation of biodiversity and wealth of the natural resources of a country, the implementation of different environmental technologies, environmental safety, the level of public health that is connected with the state of the environment, etc. An indicator that most comprehensively describes the state of the environmental field is the environmental performance index (EPI). At the same time, the environmental performance index is an indicator of the load on the environment and the rational use of natural resources.

$$N_{env} = \left(\sum_{i=1}^{3} d_{env}^{i}\right), \tag{9.3}$$

where d_{env}^{1} is the Safety & Security, d_{env}^{2} is the Natural Environment, d_{env}^{3} is the Health.

It should be noted that one of the goals and strategies for sustainable development is to reduce inequality within country and between countries. In the framework of the research, an attempt was made to find a correlation between the values of three main areas of the country's development (economic N_{econ}, social N_{soc}, and environmental N_{env}). It is suggested to consider such ratio through the values of the corresponding drivers (economic $\sum d_{econ}^{i}$, social $\sum d_{soc}^{i}$, environmental $\sum d_{env}^{i}$).

The basis for the research is the optimal development of the country. At the same time, there is equitable development of N_{econ}, N_{soc}, N_{env} areas in past, current, and future periods (Figure 9.3).

As mentioned before, the state of each area can be described by the corresponding composite indicator in this research.

Thus, there is a continuous integral interaction of three areas of the country's development. Therefore, the most appropriate is to use such a composite indicator in calculations, which would describe the growth of all areas

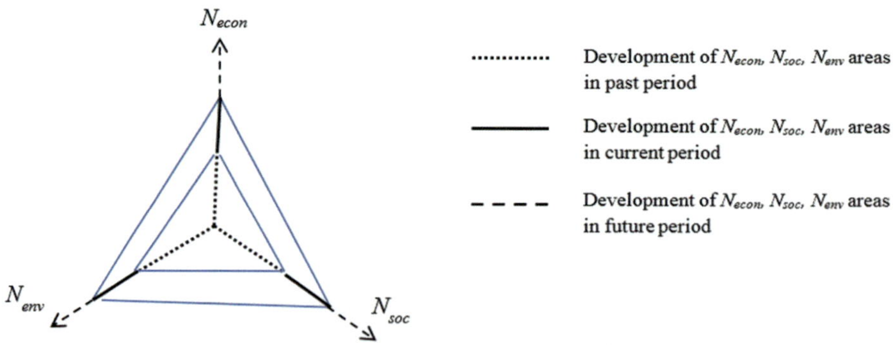

Figure 9.3 Optimal development of the country with equitable growth of economic (N_{econ}), social (N_{soc}), and environmental (N_{env}) areas of country's development.

of the country's development at the same time. The research suggests to consider an integral indicator—the Legatum prosperity index (*LPI, 2018*).

The Legatum Prosperity Index is a composite indicator that measures the achievements of the countries of the world in terms of their well-being and prosperity. It has been annually compiled since 2006 by the Legatum Institute, a British analytical center (a division of the international investment group Legatum). It is aimed at studying social well-being and its development on a global scale. The index is compiled on the basis of many different indicators, combined in 12 categories, which reflect different aspects of social life and the parameters of public welfare:

- Personal Freedom;
- Governance;
- Social Capital;
- Living Conditions;
- Education;
- Investment Environment;
- Enterprise Conditions;
- Market Access & Infrastructure;
- Economic Quality;
- Safety & Security;
- Natural Environment; and
- Health.

Thus, based on a systematic approach, the research proposes to consider LPI as a function of the economic, social, and environmental areas of the country's development. In addition, determining the relationship between the economic, social, and environmental areas of development should be a precondition of the research (9.4).

$$\begin{cases} LPI = f\left(N_{econ}, N_{soc}, N_{env}\right) \\ N_{econ} = f\left(\sum_{i=1}^{4} d_{econ}^{i}\right) \\ N_{soc} = f\left(\sum_{i=1}^{5} d_{soc}^{i}\right) \\ N_{env} = \left(\sum_{i=1}^{3} d_{env}^{i}\right) \\ N_{econ} = f\left(N_{env}\right) \\ N_{soc} = f\left(N_{econ}\right) \\ N_{env} = f\left(N_{cos}\right) \end{cases} \tag{9.4}$$

where LPI is the Legatum Prosperity Index, N_{econ} is the economic area of country's development, N_{soc} is the social area of country's development, N_{env} is the environmental area of country's development, d^i_{econ} is the economic drivers of country's development, d^i_{soc} is the social drivers of country's development, d^i_{env} is the environmental drivers of country's development.

The corresponding *LPI* values of the country and indicators of the corresponding area of the country's development are given in Table 9.1. The statistical data of 158 countries of the world for 2018 were used in the research. It should be noted that due to the methodological improvements of each next version of the indicators of the state of the country's development areas, it is impossible to use time series. In this regard, the data of one year by geographical criteria were used for calculations.

We have put forward following hypotheses in our research:

H1: Equitable development by three areas (N_{econ}, $N_{soc,}$ N_{env}) leads to the sustainable development of the country.

H2: The development of one area under condition of unequable development of others leads to the permissible inequality in country's development.

H3: The overdevelopment of one area *at the expense of other leads to the destructive inequality.*

9.4 Results

9.4.1 The Relationship of the Economic, Social, and Environmental Areas of Country's Development

Correlation and regression analysis on the dependence between LPI of a country and indicator of corresponding area of country's development based on Formula 9.4 were conducted. The correlation and regression relationship between the economic, social, and environmental areas of country's development also were identified.

The following results were obtained based on Formula 9.4 and correlation and regression analysis (Table 9.2).

The results of the analysis indicate the existence of a correlation relation between studied components. The correlation coefficients (*R*) are greater than 0.7 and are maximum close to 1. The Chaddok scale for assessment of stochastic relationships was used to estimate the tightness of relationships between the variables. According to this scale: if $R \geq 0.7$—the relationship is tight. This indicates a high degree of direct linear relation between *LPI* and the economic, social, and environmental areas of countries' development.

Table 9.1 Value of *LPI* and indicators of the economic, social, and environmental areas of country's development[1]

No.	Country ranking[2]	Country	LPI[2]	GDPpc, thous. per capita N_{econ}	HDI[3], N_{soc}	EPI[4], N_{env}
1	1	Denmark	83,96	52,12	0,93	81,60
2	2	Norway	83,96	74,35	0,95	77,49
3	3	Switzerland	83,64	64,64	0,94	87,42
4	9	Luxembourg	80,95	106,70	0,90	79,12
5	11	Great Britain	80,70	45,70	0,92	79,89
6	14	Canada	80,01	49,65	0,92	72,18
7	15	Singapore	79,15	100,345	0,93	64,23
8	16	Australia	78,64	52,37	0,93	74,12
9	17	USA	77,75	62,60	0,92	71,19
10	18	Japan	77,06	44,22	0,91	74,69
11	41	Qatar	66,45	130,47	0,84	67,8
12	69	Saudi Arabia	57,94	55,94	0,85	57,47
13	95	Ukraine	54,50	9,28	0,75	52,87
14	99	India	53,77	7,87	0,64	30,57
15	157	South Sudan	29,76	1,50	0,41	51,49

Notes and Source:

[1]The values of the indicators are calculated for 2018, since at the time of the research there were no data for subsequent years.

[2]The 2018 Legatum Prosperity Index (LPI). Ranking of 157 countries of the worlds by LPI.

[3]The 2018 Human Development Index (HDI). *Log-value is taken for calculation.*

[4]The 2018 Environmental Performance Index (EPI).

Source: Compiled by authors based on the 2018 Legatum Prosperity Index, the 2018 Human Development Index and the 2018 Environmental Performance Index to determine $N_{econ,}$ $N_{soc,}$ N_{env}.

Table 9.2 The results of correlation and regression analysis of *LPI* and the indicators of the economic, social, and environmental areas of country's development

Studied dependence $y=f(x)$	The results of correlation and regression analysis: Equation of pairwise linear regression	R	R^2
$LPI = f(N_{econ}, N_{soc}, N_{env})$	$LPI = 6{,}3 + 0{,}08\, N_{econ} + 0{,}48\, N_{soc} + 0{,}27\, N_{env}$	0,94	0,88
$LPI = f(N_{econ})$	$LPI = -56{,}93 + 1{,}35\, N_{econ}$	0,78	0,61
$LPI = f(N_{soc})$	$LPI = 2{,}97 + 0{,}76 \cdot N_{soc}$	0,91	0,84
$LPI = f(N_{env})$	$LPI = 10{,}14 + 0{,}84 \cdot$ N_env	0,86	0,74
$N_{soc} = f(N_{econ})$	$N_{soc} = -58{,}53 + 111{,}94$ *N_econ*	0,77	0,59
$N_{env} = f(N_{soc})$	$N_{env} = 0{,}15 + 0{,}01$ *N_soc*	0,84	0,71
$N_{econ} = f(N_{env})$	$N_{econ} = -47{,}04 + 1{,}21$ *N_env*	0,71	0,51

The determination coefficient shows that the variation of y is caused by the variation of x by 51–88% by studied areas of countries' development. Thus, the economic, social, and environmental areas of the country's development affect the formation of *LPI*. To confirm this, we will consider how the ranking of countries according to *LPI* will change with equitable and unequable comparison of these three areas with each other: economic N_{econ}, social N_{soc}, and environmental N_{env}. At the same time, an increase in *LPI* determines the implementation of the concept of sustainable development.

9.4.2 Equitable Development of Three Areas (N_{econ}, N_{soc}, N_{env}) Leads to Sustainable Development of the Country

The rationale for such an equitable development is proved in the research (Matvieieva et al., 2019). In particular, the research study the influence of social production on the ecological state of the country and determining the optimal ratio between the rate of development of the economic, social, and environmental areas of country's development. In the authors' opinion, the effectiveness of the country's sustainable development depends on the level of the economic, social, and environmental balance of the territory. To determine this indicator, a methodology for comprehensive diagnostics of the country's development is suggested. It is based on the index method and the method of correlation and regression analysis.

It should be noted that to determine the level of the economic, social, and environmental balance, it is proposed to use the graphical method. According to this method, the level of balance is calculated by the ratio of the areas of triangles that characterize the existing and optimal state of the economic, social, and environmental areas of country's development.

This method was used as a basis for building vector models of the country's functioning based on the ratios of the economic, social, and environmental development areas. In this regard, based on Table 9.1, the statistical indicators were recalculated into dimensionless ratios (n). The purpose of this calculation is to reduce the differently measured indicators to a single dimension and to compare their comparability. This is a precondition for determining the value of the inequality between the economic, social, and environmental areas of country's development.

The calculation was carried out by determining the maximum (N_{max}) and minimum (N_{min}) values for each of the studied indicators as well as by dividing the difference between the actual indicator for each country (N_i) shown in Table 9.1 and the minimum value by the difference obtained between the maximum and minimum value of the corresponding indicator (9.5).

$$n = \frac{N_i - N_{i\,(min)}}{N_{i\,(max)} - N_{i\,(min)}}, \tag{9.5}$$

where N_i is the actual value of ith indicator, $N_{i(min)}$ is the minimum value of indicator to be assessed, $N_{i(min)}$ is the maximum value of studied indicator.

The results of the conversion of indicators of the economic, social, and environmental areas of the countries' development into dimensionless ratios are given in Table 9.3.

According to the results obtained (Table 9.3), only few countries have the most equal ratios n (Luxembourg, Singapore). This is the basis for determining the concept of "optimal equality" in the context of a sustainable development strategy.

The research suggests considering the concept of "optimal equality" as a condition when all areas (drivers) of the country's development are equal.

For these countries, the ratios of the economic, social, and environmental areas of development range from 0.613 to 0.967. That is, there is a deviation of the indicator n within 36.6%. Table 9.2 shows that these countries have rather high LPI (from 79.15 to 80.95). Thus, the conditions for optimal equality can be represented in the form of Formula 9.6:

$$\begin{cases} 0.613 < n < 0.967 \\ 79.15 < LPI < 80.95 \\ n_{econ} = n_{soc} = n_{env} \end{cases} \tag{9.6}$$

Table 9.3 Ratios of the economic (n_{econ}), social (n_{soc}), and environmental (n_{env}) areas of countries' development

No.	Country	*LPI*	N_{econ}	N_{soc}	N_{env}
1	Denmark	83,96	0,396	0,958	0,903
2	Norway	83,96	0,567	1,000	0,834
3	Switzerland	83,64	0,493	0,986	1,000
4	Luxembourg	80,95	0,817	0,922	0,862
5	Great Britain	80,70	0,347	0,941	0,874
6	Canada	80,01	0,377	0,945	0,746
7	Singapore	79,15	0,768	0,967	0,613
8	Australia	78,64	0,398	0,972	0,778
9	USA	77,75	0,477	0,941	0,729
10	Japan	77,06	0,335	0,932	0,788
11	Qatar	66,45	1,000	0,816	0,673
12	Saudi Arabia	57,94	0,426	0,832	0,501
13	Ukraine	54,50	0,066	0,646	0,424
14	India	53,77	0,055	0,468	0,052
15	South Sudan	29,76	0,006	0,062	0,401

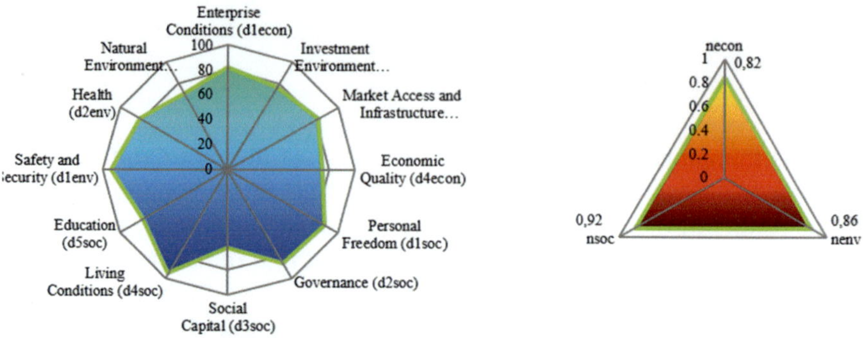

– Country development model based on the components of the *LPI*.

– Vector model based on ratios of the economic, social, and environmental areas of country's development.

Figure 9.4 Models of the development of Luxembourg under conditions of permissible inequality based on the components of the *LPI* and ratios of the economic, social, and environmental areas of country's development.
Source: Conducted by authors based on the data of the 2018 Legatum Prosperity Index.

Figure 9.4 shows the constructed models of the countries' development under conditions of optimal equality based on the data of Table 9.3 and the components of the *LPI* that is represented as drivers of the economic (d^i_{econ}), social (d^i_{soc}), and environmental (d^i_{env}) areas of country's development.

Figure 9.4 represents maximum balanced pie chart (constructed by the economic, social, and environmental drivers) and equitable vector diagram (constructed by areas of country's development). Thus, the equitable development of three areas (N_{econ}, $N_{soc,}$ N_{env}) leads to the balanced development of the country. And this, in turn, corresponds to the conditions of sustainable development.

9.4.3 The Development of One Area Under Condition of Unequable Development of Others Leads to Permissible Inequality in Country's Development

Table 9.2 shows that the highest LPIs (from 83.64 to 83.96) in countries with unequal values of ratios of the economic, social, and environmental areas of the country's development (Denmark, Norway, Switzerland). These ratios' n range from 0.3 to 1.0. That is, there is a deviation of the indicator n within 70%. In these countries, the social and environmental components are developed more at the expense of the economic one. That is, the effective development of the economy contributes to the growth in the development of the social and environmental components. Thus, the growth of one of the areas (drivers) of the country's development leads to the transformation of others. At the same time, the country is developing effectively (Denmark, Norway, Switzerland). But there are possible scenarios when countries have ranking higher than the average according to LPI (LPI = 66.45), but at the same time, on the contrary, the economic component reaches its maximum development due to environmental and social ones (Qatar). This is the basis for determining the concept of "permissible inequality" in the context of a sustainable development strategy.

The research suggests considering the concept of "permissible inequality" as a state when the development of one area (driver) under the condition of the unequable development of others leads to an acceptable level of the country's development in the context of a sustainable development strategy.

Thus, the conditions of permissible inequality can be presented as Formula 9.7.

$$\begin{cases} 0,335 < n < 1 \\ 66,45 \leq LPI < 83,96 \\ n_{econ} \neq n_{soc} \neq n_{env} \end{cases} \tag{9.7}$$

Figures 9.5 and 9.6 show the models of countries' development under conditions of permissible inequality based on the components of LPI and data of Table 9.3.

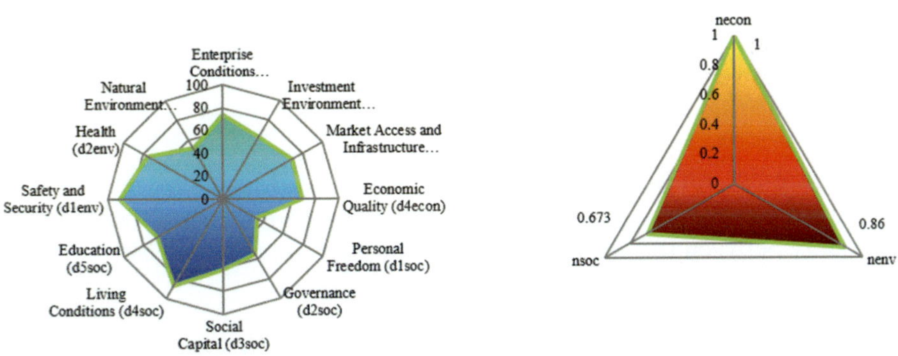

- Country development model based on the components of the *LPI*.

- Vector model based on ratios of the economic, social, and environmental areas of country's development.

Figure 9.5 Models of the development of Qatar under conditions of permissible inequality based on the components of LPI and ratios of the economic, social, and environmental areas of country's development.

Source: Conducted by authors based on the data of the 2018 Legatum Prosperity Index.

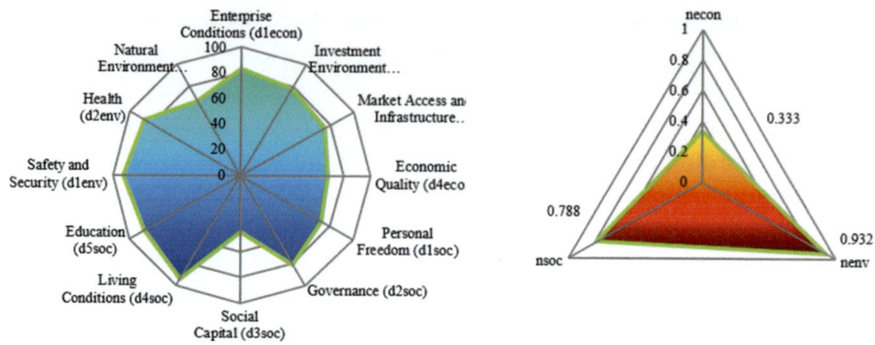

– Country development model based on the components of the *LPI.*

– Vector model based on ratios of the economic, social, and environmental areas of country's development.

Figure 9.6 Models of the development of Japan under condition of permissible inequality based on the components of the *LPI* and ratios of the economic, social, and environmental areas of country's development.
Source: Conducted by authors based on the data of the 2018 Legatum Prosperity Index.

Figures 9.5 and 9.6 show the insufficiently balanced pie chart (constructed by the economic, social, and environmental drivers) and unequable vector diagram (constructed by areas of country's development). Thus, the unequable development of one of the three areas (N_{econ}, N_{soc}, N_{env}) leads to the acceptable development of the country. This also corresponds to the conditions of the sustainable development strategy, as under conditions of optimal equality.

Therefore, in accordance with hypothesis 2, Figure 9.7 graphically presents the country's development scheme under conditions of permissible inequality between the economic, social, and environmental areas.

Case

During 2005–2010, while preparing the Development Strategy of Lublin (Poland), such areas of the economy as food industry, outsourcing, information technology, logistics, automotive industry, and biotechnology were analyzed. Labor was the key resource that was reallocated by these areas of the country's economic development. The main purpose was to determine the drivers of the country's development, under the condition of an increase in the number of jobs. Earlier it was believed that the food industry would become

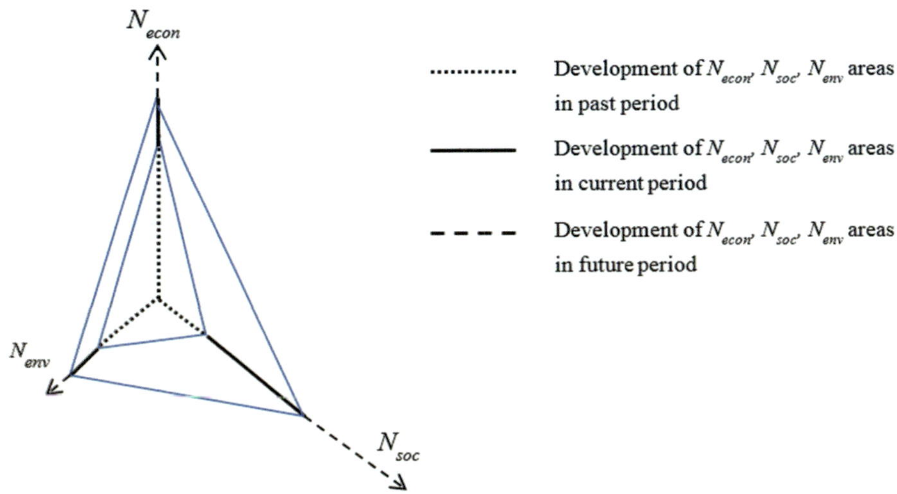

Figure 9.7 Country's development with unequable growth of economic N_{econ}, social N_{soc}, and environmental N_{env} areas.

the driver of the economic development. But it was calculated that the food industry would ensure only 15% of the employment, while outsourcing130%, information technology150%, logistics92%, the automotive industry112%, biotechnology46%. This is due to the high automation of production processes that did not provide additional job opportunities. And this, in turn, did not give additional tax revenues from the wage fund.

At that time, information technology and outsourcing attracted a large number of labor resources and became drivers for the development of the city. Source: Jastrzębska, 2010.

9.4.4 Destructive Inequality

However, the level of inequality has its limitations. Inequality of reallocation of resources can lead to both growth and decline of each country's development areas: economic, social, and environmental. It is confirmed by the analysis of ratios of the economic (n_{econ}), social (n_{soc}), and environmental (n_{env}) areas of countries' development (Table 9.2). Countries like Ukraine, India, Uganda, South Sudan and others became an example of it.

The studied indicators of these countries range from 0.006 to 0.832. That is, there is a deviation of n indicator within 99.2%. Such significant deviations of the indicators lead to a decrease in *LPI*, which is a manifestation of destructive inequality of the country's development. This is a situation when the growth of one of the drivers of the country's development leads to significant transformations of others. For example, economic growth leads to a greater extent to the environmental problems. The development of the social and environmental component leads to an increase in public expenditure, which leads to economic stagnation etc. These processes lead to significant imbalances in the countries' development in the world.

Thus, the conditions of destructive inequality can be presented in Formula 9.8.

$$\begin{cases} 0{,}006 < n < 0{,}832 \\ 29{,}76 < LPI < 66{,}45 \\ n_{econ} \neq n_{soc} \neq n_{env} \end{cases} \tag{9.8}$$

Figures 9.8 and 9.9 show the models of countries' development under conditions of destructive inequality based on *LPI* components and data of Table 9.2.

Figures 9.8 and 9.9 show unbalanced pie charts (constructed by economic, social, and environmental drivers) and unequable vector diagrams (constructed by areas of country's development). Thus, the overdevelopment of one of the three areas (N_{econ}, $N_{soc,}$ N_{env}) leads to the destructive development of the country. This is evidenced by a low rate of *LPI*. And this, in turn, does not meet the conditions of sustainable development.

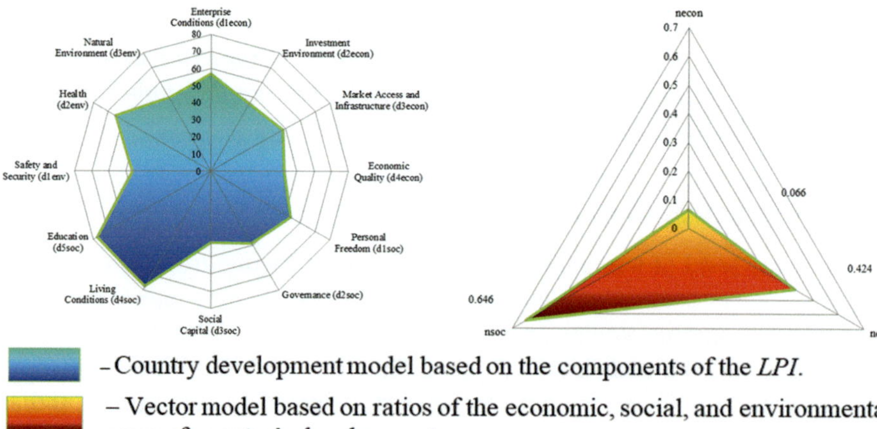

– Country development model based on the components of the *LPI*.

– Vector model based on ratios of the economic, social, and environmental areas of country's development.

Figure 9.8 Models of the development of Ukraine under conditions of destructive inequality based on the components of the *LPI* and ratios of the economic, social, and environmental areas of the country's development.
Source: Conducted by authors based on the data of the 2018 Legatum Prosperity Index.

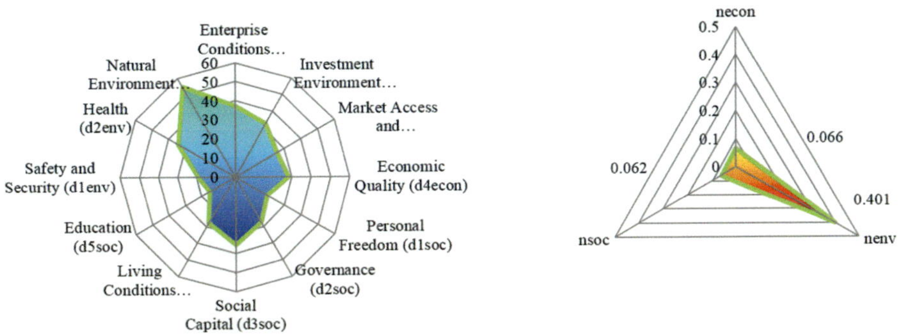

Figure 9.9 Models of the development of South Sudan under conditions of destructive inequality based on the components of *LPI* and ratios of the economic, social, and environmental areas of the country's development.
Source: Conducted by authors based on the data of the 2018 Legatum Prosperity Index.

9.4.5 Models of Countries' Development Under Condition of Growing Inequality of Resources Reallocation

According to the abovementioned, Figure 9.10 shows three models of the country's development with growing inequality while reallocating the resources.

The ranking of countries according to *LPI* allowed to determine a number of countries for which the unequable reallocation of resources by three areas (economic, social, and environmental) led to a number of positive transformations. The research analyzed the drivers for each area of countries' development (N_{econ}, N_{soc}, N_{env}). Besides, the drivers that led to the prosperity of the country in terms of implementing the concept of sustainable development were identified. Let us consider the key drivers for different countries.

Cases

The key driver of economic growth in Africa in 2012 was the increase in private consumption, followed by the amount of gross investment and government consumption (Minh, 2014).

Equitable growth (3.1%) was observed in Mauritius, where the key drivers of economic growth were the diversification of the economy with access to banking services providing and to the manufacturing sector (Zafar, 2011).

According to the results of the research of the London-based think-tank—Legatum Institute, the economic growth drivers were identified in such countries as Japan, Iceland, France, Belgium, Australia, the USA, Canada, Finland, Great Britain, Luxembourg, Singapore, Norway, Denmark, Germany, Switzerland, Sweden, the Netherlands, and New Zealand.

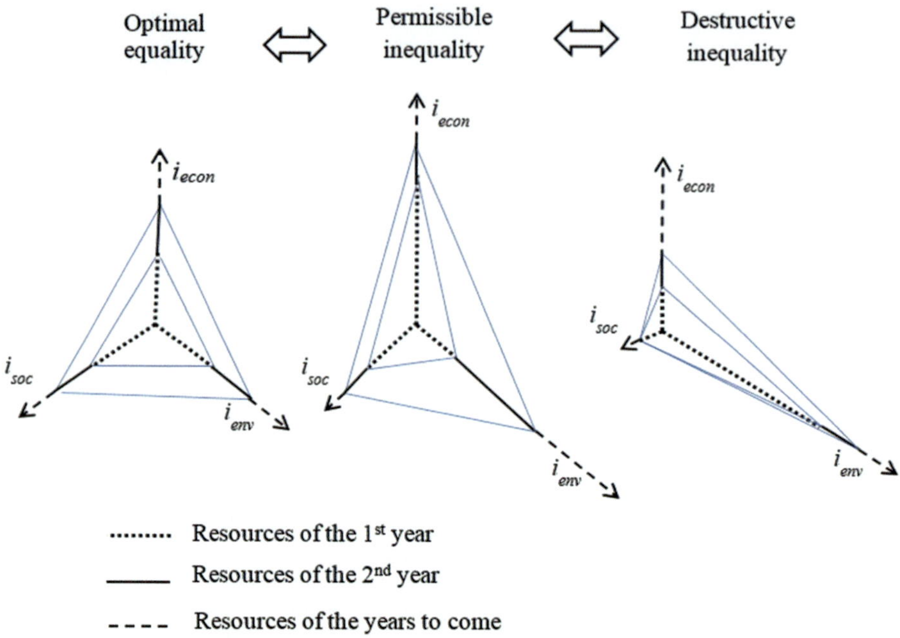

Figure 9.10 Models of the country's development with growing inequality while reallocating the resources.

The economy of Japan is one of the most developed in the world. It occupies a rather high position in the world countries ranking in terms of prosperity index by use of different innovative technologies, developed transport, energy and agricultural sectors. Japan ranks third in terms of GDP and industrial output, seconds only to the USA and China.

In Iceland, the drivers of economic growth are the intensive diversification of industries based on cheap renewable energy, the development of biotechnology and the tourism industry. In addition, fishing and fish processing are the basis of the economy. This type of activity makes 12% of GDP.

France ranks second in the European Union in terms of the total amount of economy. It is possible to identify such drivers of the country's economic development as the service sector, which employs 76.7% of the economically active population as well as has high rates in tourism, energy, transport and agriculture.

Drivers of economic growth in Belgium include the implementation of different innovative technologies (in the areas of telecommunications, microelectronics, biotechnology, etc.) into the country's economy in order to uplift the national economy to the level of world standards.

The economy of Australia is one of the largest economies in the world with GDP of 15.7 trillion US dollars. The economy of Australia depends on commodities. Iron ore is one of the country's most important resources. The service sector dominates in the economy of Australia and covers about 68% of GDP. Economic growth depends on the mining and agricultural sectors, whose products are mainly exported to East Asian markets.

The key feature of the US economy is its diversification (from lat. diversification – changing, diversity). In addition, the US economy stands out with the highest nominal value of external debt in the world. From a global point of view, it occupies high positions, but a huge difference in the incomes of the population is a problem to be solved by the government in the near future.

The key role in creating a favorable economic situation in Canada was played by the diversity and wealth of the country's natural resources. Canada, more than other countries, directly depends on its oil reserves. Due to the influence of various factors that prevented oil production, the level of GDP was rapidly falling. Despite everything, Canada continues its policy of attracting entrepreneurs and expanding the scope of activities.

The experts consider that the key drivers of economic growth in Finland are political stability, an open economic sector, highly developed infrastructure, reliable telecommunications as well as a high level of interaction between enterprises, scientific and technical centers and universities, mastering new technologies, readiness for changes, high educational level of the population, and favorable business climate.

Austria belongs to the leaders of economic development of the European Union countries. Almost all sectors of the economy are developed in the country, from tourism to aerospace technologies. Drivers of economic growth include the active use, implementation of innovative technologies and new equipment into all fields of activity.

The driver of economic growth in Great Britain is the service sector (3/4 of GDP). It should be noted that Great Britain provides 10% of the world export of services, namely banking, insurance, brokerage, consulting, as well as in the field of computer programming. Similar drivers of economic growth are observed in Luxembourg. Services account for more than 85% of the country's GDP.

It should be noted that Luxembourg is heading toward sustainable development; as a result, different environmental technologies are actively implemented in the country. The development of innovative clusters also plays an important role, especially in the field of biomedicine, environmental technology, space, and information technology.

The key drivers of economic growth in Singapore are transnational corporations, freedom from corruption. At the same time, the economic growth of the country depends on the export of products, especially in such fields as consumer electronics, information technology, pharmaceuticals, and financial services.

The main driver of economic prosperity in Norway is the export of raw materials and semi-finished goods, as Norway can be proud of its natural resources, such as hydropower, forest, fish, minerals, etc.

The basis of the economy of Denmark is the services sector. The developed market economy of the country depends on imported raw materials and foreign trade due to own small reserves of natural resources. Denmark regulates the economic activity and inflation through fiscal policies aimed at supporting the exchange rate to guarantee the stability of the national currency.

The economy of Germany is the largest economy in Europe. It is characterized by a well-developed infrastructure and skilled labor. Therefore, one of the drivers of social development of the economy is the qualification and specialization of labor potential. The country is known worldwide as the largest exporter of machinery, vehicles, chemicals and household equipment. Germany is a post-industrial country. The basis of its economy (from 70% to 78% depending on the period) consists of services, 23–28%—production and 5–15%—agrobusiness.

The key drivers of economic growth in Switzerland are industry, trade and the fact that investors are sure that their funds are safe.

The economy of Sweden is developed and export-oriented one. The basis for economic growth is environmental drivers, namely: forest, hydropower, and iron ore. The country has a modern infrastructure, excellent internal and external communications, and a highly skilled labor.

At the same time, in the Netherlands, the economic growth is caused by environmental drivers. Moreover, they have a positive impact on the social and economic fields. Thus, the Netherlands is known all over the world for its flowery fields that provide export of products and ensure a huge number of jobs not only for the Dutch, but also for foreigners. In 2018, the country's nominal value of GDP exceeded 913 billion US dollars. The Netherlands occupies rather high position in the global ranking of countries in terms of prosperity.

New Zealand is the country with the most stable, balanced, and open economy. An interesting fact is that the success of the country's economy directly depends on the dairy industry: in particular, New Zealand company Fonterra, provides dairy products for more than 30% of the world.

Source: Minh (2014), Zafar (2011), online resource *Uteka.ua* (2016).

9.5 Conclusions and Discussion

Based on the study of 157 countries of the world, the economic, social, and environmental areas of the countries' development are analyzed, and the level of inequality and the associated consequences are identified. Three country's development models are suggested based on *LPI*. These models are constructed based on an assessment of the inequality level, which allowed forming a rating of countries according to the value of harmonization of the economic, social, and environmental areas of the country's development and their drivers.

Based on the correlation and regression analysis, the relationship between the studied areas of the country's development as well as between them and LPI is proved.

The analysis of multiple regression $LPI = 6{,}3 + 0{,}08\ N_{econ} + 0{,}48\ N_{env}$ proves that different coefficients of variables indicate the fact that among countries the unequable development between the economic, social, and environmental areas dominates. At the same time, the social component has the maximum impact on *LPI*.

The classification of countries by the level of inequality between economic, social, and environmental areas of development based on *LPI* is suggested. According to the classification, there are countries with optimal equality, permissible inequality, and destructive inequality. For developing countries under conditions of sustainable development, *LPI* is in the range from 66.45 to 83.96, and for countries with destructive inequality, the *LPI* is in the range from 29.76 to 66.45. Thus, the equality between the economic, social, and environmental areas of the country's development and the permissible inequality between them within 70% lead to an increase in LPI, and, accordingly, to the country's prosperity in the context of the implementation of the sustainable development strategy. At the same time, the growth of inequality between these areas by more than 70% leads to a destructive transformation in the process of country's development.

References

1. Ali, Zafar (2011) Mauritius: An Economic Success Story. Yes, Africa can: success stories from a dynamic continent / editor: Punam Chuhan-Pole, Manka Angwafo. Retrieved from: http://documents.worldbank.org/curated/en/304221468001788072/pdf/634310PUB0Yes0061512B09780821387450.pdf [Online Resource].
2. Basiago, A.D. (1999). Economic, social, and environmental sustainability in development theory and urban planning practice. *Environmentalist*, 19, 145–161. https://doi.org/10.1023/A: 1006697118620.

3. Brundtland, G.H. (1987). Our common future: Report of the World Commission on environment and development. Geneva, UN-Dokument A/42/427. Retrieved from: http://www.un-documents.net/ocf-ov.htm [Online Resource].
4. Dao, Minh. (2014). Drivers of Economic Growth in Developing Countries. *Journal for Studies in Economics and Econometrics*, 38, 75–85.
5. Dechert, W.D., Nishimura, K. (1983) A complete characterization of optimal growth paths in an aggregated model with a non-concave production function. *Journal of Economic Theory*, Vol. 31(2), 332–354.
6. Edward B. Barbier and Joanne C. Burgess (2017). The Sustainable Development Goals and the systems approach to sustainability. *Economics Discussion Papers*, No 2017-28, Kiel Institute for the World Economy. Retrieved from: http://www.economics-ejournal.org/economics/discussionpapers/2017-28 [Online Resource].
7. Greiner, A. (2007). The dynamic behavior of an endogenous growth model with public capital and pollution. *Studies in Nonlinear Dynamics&Econometrics*, 11(2), 1375–1375. https://doi.org/10.2139/ssrn.434020.
8. IMF (2020). GDP per capita 2018. Retrieved from: https://www.imf.org/external/pubs/ft/weo/2019/02/ [Online Resource].
9. Jastrzębska Anna (2010). Systematic solutions for Lublin and Ukrainian cities in the field of creating modern self-government / Editor Anna Jastrzębska / City Hall of Lublin 2010, 184 p. Retrieved from: http://www.mvk.if.ua/uploads/files/anons_231110.pdf [Online Resource].
10. Matvieieva, Y., Myroshnychenko, I., Valenkevych, L. (2019). Optimization Model of the Socio-Ecological-Economic Development of the Administrative Territory. *Journal of Environmental Management and Tourism*, Vol. 8 (40), 1888 - 1899. https://dx.doi.org/10.14505/jemt.v10.8(40).18].
11. Munasinghe, M. (1992). Environmental economics and sustainable development. The UN Earth Summit, Rio de Janeiro, Environment Paper No.3, World Bank, Wash. DC, USA.
12. Munasinghe, M. (1994). Sustainomics: a transdisciplinary framework for sustainable development. Keynote Paper, Proc. 50th Anniversary Sessions of the SriLanka Assoc. for the Adv. Of Science (SLAAS), Colombo, SriLanka.
13. Pavlenko,O., Martynets, V., Dreval, O., & Smolennikov D. (2020) Analysis of Influence of the Quality of Specialist Training on Social and Economic Development. *Quality*, Vol. 21 (176), 81–86.

14. PROSPERITY (2018). The 2018 Legatum Prosperity Index. Retrieved from: https://www.prosperity.com/rankings?pinned=&filter=; [Online Resource].
15. Purvis, B., Mao, Y. and Robinson, D. (2019). Three pillars of sustainability: in search of conceptual origins. *SustainSci*, 14, 681–695. https://doi.org/10.1007/s11625-018-0627-5.
16. Stachurski J., Venditti A., Yano M. (Eds.). (2012). Nonlinear Dynamics in equilibrium models: chaos, cycles and indeterminacy. Berlin: Springer. 127–150.
17. Telizhenko, O., Pavlenko, O., Martynets, V. and Rybalchenko, S. (2019). Modeling the influence of cluster components on the economic development of a territory. *TEM Journal*, 8(3), 900-907. https://dx.doi.org/10.18421/TEM83-30
18. UN General Assembly (2015). Transforming our world: the 2030 Agenda for Sustainable Development. Retrieved from: https://www.refworld.org/docid/57b6e3e44.html [Online Resource].
19. UNDP (2018). The 2018 Human Development Index. Retrieved from: http://hdr.undp.org/ [Online Resource].
20. UTEKA (2020). Top 19 economically developed countries of the world. Retrieved from: https://uteka.ua/ua/publication/news-14-delovye-novosti-36-top-19-ekonomicheski-razvityx-stran-mira) [Online Resource].
21. Wendling, Z. A., Emerson, J. W., Esty, D. C., Levy, M. A., de Sherbinin, A., et al. (2018). 2018 Environmental Performance Index. New Haven, CT: Yale Center for Environmental Law & Policy. Retrieved from: https://epi.yale.edu/ [Online Resource].

10

Financial Inequality in Territorial Communities: International Experience and Case-Study of Ukraine

Yuriy Petrushenko[1], Anna Vorontsova[2], Oksana Ponomarenko[3]

[1,2,3]Department of International Economic Relations, Sumy State University, Ukraine

ABSTRACT

Decentralization of power is a prevalent trend in the modern world that demonstrates the significant popularity of the idea to solve urgent problems of communities most efficiently at the local level. It also provides for local authorities' resource provision and autonomy in solving the main social problems of the local population and is used as an effective instrument for regional development. Ukraine also made several political steps to implement the concept of decentralization, which contributed to a large-scale administrative-territorial reform that is taking place to date. These moves are definitely positive achievements, but there are some difficulties in their implementation. One of the main difficulties between the new administrative entities in the united territorial communities is the emergence of financial inequality due to their different financial capacity. The main purpose of the research is to develop its own approach for estimation of the united territorial communities' financial capacity in Ukraine as an indicator of their financial inequality, based on the analysis of the current domestic and international methodologies. For this reason, the main stages and achievements of the decentralization process in the world and Ukraine were analyzed and the shortcomings of the existing methodologies for estimation of the united territorial communities' financial capacity were identified. As a result, the authors proposed their methodology of conducting an integral estimation of the united territorial communities' financial capacity, based on the taxonomic method for all united territorial communities of Ukraine created in 2016–2018 using the MS Excel toolkit. The obtained results were distributed, using the modified Harrington scale, which allowed to sort the studied subjects into five groups of financial inequality according to their financial capacity level: crisis, low, satisfactory, moderate (above average), and high. The study results reveal that new administrative entities have especially low-capacity levels, which can be increased due to appropriate actions of the government.

10.1 Introduction

Ukraine has officially acknowledged the eurointegration vector of its development that causes radical changes in most sectors of the economy. One of the most important and fundamental reforms in this regard is the decentralization of power that provides a real strengthening of local authorities and also ensures the separate territorial units (communities) with sufficient resources to deal with their development problems. These goals are declared in the Ukraine-2020 Sustainable Development Strategy, as well as in the State Strategy for Regional Development for the period up to 2020, and envisage a structural adjustment not only in economic relations but also in other related sectors of the national economy (for example, in education, health care, etc.) The goal of the government at this stage of the country's development is to create a territorial system of local self-government that will be able to ensure the sustainable development of its territories and to fulfill a wide range of received powers.

In this concern, the topic of research is highly relevant and requires a thorough analysis of how to build an effective financial management system of territorial communities and for this purpose, the methodology of the estimation of UTC financial capacity needs to be studied. This issue is important at the level of regional state administrations and regional councils for preparation and adoption of perspective plans during the community territories formation and it is also essential at the level of the communities themselves.

10.2 Literature Review and Bibliometric Analysis

Summarizing the European experience, it can be concluded that most countries have passed or are in the process of reforming their local self-government systems regarding decentralization. The main provisions of this process were ratified in the form of the European Charter of Local Self-Government, which was adopted by Ukraine in 1998. In this regard, numerous works of scientists were devoted to the study of foreign experience is the process of decentralization and possible options for its implementation for Ukraine. This topic was the field of research for Pygul N.G. and Luta O.V. (2016), Kasich A.O. and Petrushko A.S. (2017), Udovychenko V., Melnychuk A., Gnatiuk O., Ostapenko P (Udovychenko et al., 2017), Pyankovsky V. (2018) and many others.

In the same context, it is appropriate to cite studies, devoted to move toward a model of decentralization and its impact on the socio-economic and innovative development of the country: in particular, Goel R., Mazhar U., Nelson M., Ram R. (Goel et al. 2017), Chygryn O., Petrushenko Y. and oth. (Chygryn et al., 2018), Vasylieva T., Harust Yu., Vynnychenko N. and oth. (Vasylieva et al., 2018), Melnyk, L., Sineviciene L., Lyulyov O., and others. (Melnyk et al., 2018) and others.

The numerous studies in this direction have begun due to the active phase of the decentralization reform in Ukraine and the change in its administrative-territorial structure. Thus, the question of the capacity of new administrative-territorial units—territorial communities in the context of their united was the subject of research for such scholars as Zelinsky S. E. (2016), Sember S.V., Chubar O.G., Mashiko K.S. (Sember et al., 2015), Tkachenko Yu.V., Kovalchuk Ya.O. (Tkachenko and Kovalchuk, 2018), Tulai, O., Petrushenko, Y., Glova, J., Sydor, I. (Tulai et al., 2019) and others.

Currently, the financial capacity of UTC is calculated by the Expert Group for Financial Monitoring of the Central Reform Office under the Ministry of Regional Development (with the support of the U-LEAD Europe Program and the SKL International project) based on a rating approach. It is their monitoring that allows us to obtain statistical information about the main financial indicators of territorial communities and is used as a basis for our research.

According to this methodology, the financial capacity of each united territorial community is estimated by four main criteria: own revenues per one inhabitant, subsidy level of budgets, the share of expenditure on maintenance of management in own resources of UTC and capital expenditures per one inhabitant.

Speaking about the advantages of the current estimation methodology of financial capacity, it should be admitted that, it includes indicators of budgetary capacity; indicator, which identifies the management of self-government authorities; an indicator that shows the financial capacity of households; and capital expenditures that characterizes the funding of investment and innovative activities.

It is also necessary to highlight the significant disadvantage of this methodology that prevents the local self-governments from a comprehensive estimation of the financial capacity of their territorial community, its trends and taking appropriate management decisions. The fact is that the estimation of the UTC financial capacity was performed based on a rating approach. In 2016 and 2017, the experts formed the overall unified rating of UTC's financial capacity, but in 2018, the ranking approach was changed. Now all UTC are divided into four groups according to the population size criteria, and the rating is made up of each individual group. This approach is based on the hypothesis that communities with a larger population are more capable.

Thus, the research goal includes an analysis of current methodology and development of its own approach of the financial capacity estimation of UTC in Ukraine, as the indicator of their financial inequality.

The paper is organized as follows. Section 1 is devoted to the brief historical review of the peculiarities of the decentralization reform in Ukraine. Section 2 is devoted to the description of the methods proposed by the authors for conducting an integrated evaluation of the financial capacity of the UTC based on the taxonomic method and its results. The conclusions summarize the obtained results and provide the basis for further research of authors in this field of research.

10.3 Brief Historical Excursion for Decentralization in the World and in Ukraine

For developed countries, the decentralization of functions and authorities, as well as financial resources between the central government and the regions, is the basis for the state development and the well-being of its citizens. The efficiency of this mechanism can be clearly seen in the example of both post-socialist countries of Europe (Poland, Lithuania, and Estonia) and developed European democracies (Sweden, Great Britain, Spain, France, Germany, etc.).

Antoni Castells (2001) explores intergovernmental finance issues in the process of decentralization in Spain. It is emphasized that financial decentralization, as a part of political decentralization, should achieve two political goals: autonomy and equality.

However, both wealthy and poor regions are concerned about their financial interests. The experience of decentralization shows us, that rich regions believe they will make excessive fiscal transfers to poor regions, but in fact, rich regions clearly benefit from the capital effect. There are also cases where poor regions expect that financial decentralization will have a negative impact on them as they become dependent on their own financial capacity.

Obviously, this situation is common to the current reality of Ukraine. In this way, financing autonomous government has become a critical issue with important financial and political implications for Spain. The current system is mainly based on grants from the central government, while tax revenues are inadequate. Grants make up 72.80% of the autonomous government's revenue, and they come from the central government at 63.44%. Tax revenue is 25.90% of local resources. In Spain, the future evolution of the intergovernmental finance system aimed at giving more fiscal responsibility and transparency to regional governments, reforming the grant system and improving institutional coordination mechanisms, which is relevant to the situation in Ukraine.

For a more complete picture of the impact of financial decentralization on regional development, the experience of China was examined. Thus, Kanbur and Zhang (2001) studied the effect of decentralization on inequality. They have empirically proved that fiscal decentralization and trade liberalization also systematically influence regional inequality. The study suggested that decentralization affects regional inequalities during the economic transition from a planned to a market economy. Both, decentralization and openness to global markets, provide local governments with the opportunity to make better use of comparative advantages. These conclusions could be beneficial to Ukraine.

Since April 2014, the government of Ukraine began to actively implement measures of administrative-territorial reform toward its decentralization. Then the concept of reforming local self-government and the territorial structure of power was approved, as well as an Action Plan for its implementation. They identified the main economic, environmental, social, and political problems that exist today in this sphere of activities in the country (primarily resource constraints that causes the poor quality of services, material depreciation of equipment, the inconsistency of the main tasks from the central authorities with the interests of society, etc.) and the main ways of their solution.

In 2015, a number of regulations were adopted, in particular, the Law of Ukraine "On Voluntary Amalgamation of Territorial Communities" and the Methodology for the formation of capable territorial communities. For the period 2015–2018, 878 united territorial communities were established in Ukraine, bringing together 4016 communities. Thus, 37.6% of the area of Ukraine is covered by the reform of the administrative-territorial structure, and 25.5% of the population lives in the territories of UTC.

In the framework of the financial decentralization, changes were made to the Budget and Tax Codes of Ukraine, and the newly formed communities were endowed with resources as cities of regional significance. According to the results of financial decentralization, the share of local budgets (with transfers) in the consolidated budget of Ukraine increased from 45.6% in 2015 to 51.5% in 2018. The newly formed communities received an expanded range of powers that intensified problems of financial capacity and financial resources of communities.

10.4 Method of Conducting Integral Estimation of the United Territorial Communities Financial Capacity as Indicator of their Financial Inequality

After analyzing and summarizing the opinions of scientists and experts in this field, we decided to use the taxonomic method for calculating the integral indicator to evaluate the financial capacity of the united territorial communities of Ukraine. Note that all calculations were done using the MS Excel toolkit.

The taxonomic method is one of the most effective tools of multicriteria analysis, which, as a result of a series of successive stages, involves the calculation of a generalization indicator based on a comparison of actual indicators with the corresponding characteristics of a certain benchmark. One of the first scientists who proposed it as a method of research is Z. Helving (1972), the key principles of which method have been used in this paper.

The algorithm for estimation of the UTC financial capacity contains the following steps (Figure 10.1):

The integral indicator of the level of financial capacity is a synthetic value, which means it includes the "equilibrium" value of all the basic indicators that characterize the general state of development of regions.

Let us consider in more detail each step of Figure 10.1.

First of all, it is necessary to form the information base of the research, on the basis of which indicators, that directly or indirectly affect the level of efficiency of the financial process in the territorial communities, will be evaluated. Thereafter we submit them in the form of the input matrix. From a mathematical point of view, such a matrix of observations will have the following form:

$$K = \begin{pmatrix} k_{11} & \dots & k_{1j} & \dots & k_{1n} \\ \dots & \dots & \dots & \dots & \dots \\ k_{i1} & \dots & k_{ij} & \dots & k_{in} \\ \dots & \dots & \dots & \dots & \dots \\ k_{m1} & \dots & k_{mj} & \dots & k_{mn} \end{pmatrix} \tag{10.1}$$

METHOD OF INTEGRAL ESTIMATION OF THE UNITED TERRITORIAL COMMUNITIES FINANCIAL CAPACITY AS INDICATOR OF THEIR FINANCIAL INEQUALITY

1 STAGE. Formation of observation matrix (K) and differentiation of indicators for incentives and disincentives

Input parameters:	
– own revenues per one inhabitant, UAH per person (*Rev*);	Incentive
– subsidy level of budgets (share of base / reverse grant in income),% (*SubsLev*);	Disincentive
– the share of expenditure on maintenance of management in own resources of united territorial communities (excluding transfers from the state budget),% (*ManagExp*);	Incentive
– capital expenditures per one inhabitant (excluding subventions from the state budget), UAH per person (*CapExp*).	Disincentive

↓

2 STAGE. Standardization of indicators of the observation matrix (formation of matrix of standardized indicators - L)

↓

3 STAGE. Formation of matrix of reference values (L_{et})

↓

4 STAGE. Calculation of quasi-distances (KV_j)

↓

5 STAGE. Calculation of the taxonomic integral indicator of the financial capacity of united territorial communities (FC_i)

Figure 10.1 Main stages of the integral estimation of the UTC' financial capacity as indicator of their financial inequality.
Source: Compiled by authors.

where K – matrix of input financial indicators;
i – the number of the relevant financial indicator, $i = 1 \div m$;
j – the number of the relevant subject (territorial community), $j = 1 \div n$;
k_{ij} – i-th financial indicator for j-th entity (territorial community)

The input data were obtained from official statistical information provided by experts from the Financial Monitoring Group of the Central Office of Reforms under the Ministry of Regional Development, Construction and Housing and Communal Services of Ukraine (supported by the U-LEAD Europe Program and SKL International). For the convenience of conducting further calculations, we will give them short variable names and differentiate by the direction of impact on incentives and disincentives (Table 10.1). Such a division is the basis for the future construction of a vector-standard.

Table 10.1 Main characteristics of the input parameters for the UTC' financial capacity estimation

Variable name	Characteristic	The direction of impact
Rev	Own revenues per one inhabitant, UAH per person	Incentive
SubsLev	Subsidy level of budgets (share of base / reverse grant in income),%	Disincentive
CapExp	Capital expenditures per one inhabitant (excluding subventions from the state budget), UAH per person	Incentive
ManagExp	The share of expenditure on maintenance of management in own resources of united territorial communities (excluding transfers from the state budget),%	Disincentive

These indicators were selected in three time periods: 2016, 2018, and 2019, in which there was a gradual development of the decentralization reform in Ukraine. Also, in this time period, it was envisaged the formation of united territorial communities. As a result, their number of observations for each analyzed year was different—159, 366, and 664, respectively. Below is a descriptive statistics on the array of input data in Table 10.2.

As we can see from the table above, with the increase in the number of UTC, the average own revenues per one inhabitant increased from 2261.08 UAH per person in 2016 to 3884.27 UAH per person in 2018. At the same time, capital expenditures per one inhabitant grew from 669.46 UAH per person in 2016 to 1337.98 UAH per person in 2018. Instead, the level of subsidization of budgets fell on average (from 14% to 11%, respectively), while expenditures on maintenance of the management in their own UTC resources grew (from 19% to 29% on average in 2016 and 2018, respectively). Indicators of asymmetry indicate the nature of the distribution, which is conventionally normal only for the subsidy level of budgets and the cost of maintaining the management. Other calculated parameters in Table 10.2 will be used in a subsequent study at selected stages of the assessment of the UTC financial capacity.

The next step is the standardization of input data. Since indicators of the model of UTC financial capability have different natures and non-dimensional values, the rationing of the elements of the matrix is carried out. To do this, the original data should be replaced by standardized, using the following formulas (Plyuta, 1980):

Table 10.2 Descriptive statistics for the set of input data for the UTC' financial capacity estimation

Variable	Observations	Mean	Standard deviation	Minimum	Maximum	Excess	Asymmetry
2016							
Rev	159	2261.08	1948.83	378.44	15997.25	18.51	3.61
SubsLev	159	0.14	0.16	-0.16	0.57	-0.09	0.61
CapExp	159	669.46	1208.54	13.40	12600.06	62.24	6.99
ManagExp	159	0.19	0.08	0.04	0.44	0.16	0.62
2017							
Rev	366	2974.00	1968.15	535.40	21524.47	26.31	3.84
SubsLev	366	0.12	0.13	-0.26	0.53	0.55	0.65
CapExp	366	1453.31	1275.80	262.57	12661.77	32.10	4.82
ManagExp	366	0.26	0.11	0.05	0.75	1.68	1.10
2018							
Rev	664	3884.27	2672.11	597.25	25501.60	17.60	3.33
SubsLev	664	0.11	0.15	-0.28	0.60	0.51	0.69
CapExp	664	1337.98	1180.41	202.34	11559.60	19.82	3.74
ManagExp	664	0.29	0.12	0.07	0.87	2.87	1.31

$$l_{ij} = \frac{k_{ij} - \bar{k}_i}{\sigma_i} = \frac{k_{ij} - \frac{1}{m}\sum_{j=1}^{n} k_{ij}}{\sqrt{\frac{1}{m}\sum_{j=1}^{n}(k_{ij} - \bar{k}_i)^2}} \tag{10.2}$$

where $\bar{k}_i$ – the arithmetic means of the corresponding indicator k_{ij};
σ_i – standard deviation of the indicator k_{ij};
l_{ij} – standardized value indicator k_{ij}.

As a result, the matrix of standardized indicators will have the following form (Plyuta, 1980):

$$L = \begin{pmatrix} l_{11} & \dots & l_{1j} & \dots & l_{1n} \\ \dots & \dots & \dots & \dots & \dots \\ l_{i1} & \dots & l_{ij} & \dots & l_{in} \\ \dots & \dots & \dots & \dots & \dots \\ l_{m1} & \dots & l_{mj} & \dots & l_{mn} \end{pmatrix} \tag{10.3}$$

where L – matrix of standardized financial indicators.

At the third stage, the formation of a vector of reference values of united territorial communities' financial capacity indicators occurs (Plyuta, 1980):

$$L_{et} = \begin{pmatrix} l_{1et} \\ \dots \\ l_{iet} \\ \dots \\ l_{met} \end{pmatrix} = \begin{pmatrix} \begin{bmatrix} \max(k_{1j}) \\ \min(k_{1j}) \end{bmatrix} + \sigma_1 \\ \begin{bmatrix} \max(k_{1i}) \\ \min(k_{1i}) \end{bmatrix} + \sigma_i \\ \begin{bmatrix} \max(k_{1n}) \\ \min(k_{1n}) \end{bmatrix} + \sigma_n \end{pmatrix} \tag{10.4}$$

where L_{et} – matrix of reference values of united territorial community's financial capacity indicators

The obtained reference values are compared with the standardized ones, which is the basis for the calculation of quasi-distances. Such procedures are carried out according to the following formula (Plyuta, 1980):

$$KV_j = \sqrt{\sum_{i=1}^{m} (l_{ij} - l_{iet})^2} \tag{10.5}$$

where KV_j – quasi-distance between the j-th and the reference entity

The final final integral indicator of the financial capacity of united territorial communities is calculating in the next way (Plyuta, 1980):

$$FC_i = 1 - \frac{KV_j}{c_0} = 1 - \frac{KV_j}{\overline{KV_0} + 2S_0} = 1 - \frac{KV_j}{\frac{1}{m}\sum_{i=1}^{m} KV_j + 2\sqrt{\frac{1}{m}\sum_{i-1}^{m}(KV_j - \overline{KV_0})^2}} \tag{10.6}$$

where FC_i – integral taxonomic index of the financial capacity of united territorial communities

On the basis of the received calculations, we propose to differentiate the obtained values of estimation of financial capacity of united territorial communities for levels of their financial inequality: crisis, low, satisfactory, moderate (above average), and high levels. Considering the peculiarities of constructing a taxonomic index whose value ranges from 0 to 1, we suggest using Harrington's scale (Harrington, 1965) to differentiate the level of financial capability of united territorial communities, modifying it according to the needs of our study.

The general structure of the distribution of united territorial communities by the level of financial capacity as an indicator of their financial inequality is shown in Figure 10.2.

As we can see, the distribution structure of the UTC of Ukraine in 2016–2018 has changed in a positive direction, which may indicate a gradual improvement of the situation regarding the financial capacity of the territorial communities. The percentage of subjects located at the crisis level of financial capacity for the analyzed period decreased by 17.6%. In our opinion, the newly formed communities fall into the crisis level group, thereafter, they gradually improve their own performance. Thus, indicators with a low and satisfactory level of financial capacity are gradually increasing.

A more detailed result of the delimitation of the UTC of Ukraine for 2016–2018 by the levels of financial capacity as an indicator of their financial inequality with details within the regions is as follows (Table 10.3).

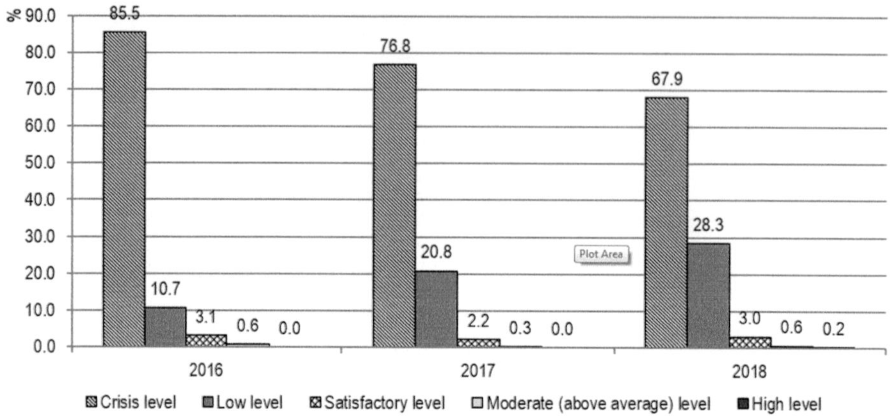

Figure 10.2 Distribution structure of UTC of Ukraine by the level of indicator of financial capacity as an indicator of their financial inequality in 2016–2018.
Source: Compiled by authors.

It should be noted that the first community whose indicators allowed it to be in a group with a high level of financial capacity, appeared in 2018. It was Slobozhanskaya UTC in the Dnipropetrovsk region. The community was created in October 2015, in 2016 it was at a satisfactory level, in 2017 it moved to the group with a moderate level, and in 2018 it has reached a high level of financial capacity. This example demonstrates the effectiveness of local self-governments' activities in using the financial resources, as well as the gradual positive effect of the decentralization reform, as itself.

It is necessary to admit that the newly formed territorial communities, as a rule, fall into the group with the crisis level. Thus, among the 22 united territorial communities of the Dnipropetrovsk region, established in 2017, 14 were at a crisis level in 2018. And this is an objectively normal process of the territorial communities' formation.

Part of the UTC, which were in a group with a crisis level in 2016, have already moved to a higher level of financial capacity in 2017 (as an example, Studenyanska, Smoligovsky, Zelenodolsk, Vakulivska UTC), however, there are some communities that remained at the same level without a positive trend. Local authorities should pay attention and optimize financial capacity indicators by adopting appropriate management decisions.

Using the Prishib community as an example, we see that in the first year after establishment, the community had a moderate level of financial capacity, however, in the next two years it moved to a lower level—satisfactory, but this is not a critical change in the community's position. Grechanopodivska

Table 10.3 Distribution results of UTC of Ukraine by levels of financial capacity as indicator of their financial inequality for 2016, 2017, and 2018

The level of financial capacity	Indicator bounds	United territorial communities by regions		
		2016	2017	2018
Crisis level	0.00–0.20	Vinnytska (1); Volynska: (5); Dnipropetrovska: (9); Donetska: (1); Zhytomyrska: (9); Zakarpat·ska: (1); Zaporizka: (6) Iv.-Frankivska: (3); Kyyivska: (1); Kirovohradska: (2) Luhanska: (2); Lvivska: (15); Mykolayivska: (1); Odeska: (6); Poltavska: (7); Rivnenska: (5); Sumska: (1); Ternopilska: (25); Khersonska: (1); Khmelnytska: (21); Cherkaska: (3); Chernivetska: (10); Chernihivska: (3)	Vinnytska: (15); Volynska: (13); Dnipropetrovska: (20); Donetska: (2); Zhytomyrska: (27); Zakarpat·ska: (2); Zaporizka: (8); Iv.-Frankivska: (11) Kyyivska: (2); Kirovohradska: (4); Luhanska: (2); Lvivska: (20) Mykolayivska: (10), Odeska: Balt·(8); Poltavska: (8); Rivnenska: (16); Sumska: (13); Ternopilska: (35); Kharkivska: (3); Khersonska (9); Khmelnytska: (21); Cherkaska: (5); Chernivetska: (16) Chernihivska: (10)	Vinnytska: (19) Volynska: (30) Dnipropetrovska: (37) Donetska: (2) Zhytomyrska: (36) Zakarpat·ska: (4) Zaporizka: (25) Iv.-Frankivska: (22) Kyyivska: (5) Kirovohradska: (7) Luhanska: (6) Lvivska: (25) Mykolayivska: (15) Odeska: (8) Poltavska (12) Rivnenska: (21) Sumska: (16) Ternopilska: (38) Kharivska: (2) Khersonska: (20) Khmelnytska: (30) Cherkaska: (17) Chernivetska: (26) Chernihivska: (27)

Table 10.3 *Continued*

The level of financial capacity	Indicator bounds	United territorial communities by regions		
		2016	**2017**	**2018**
Low level	0.21–0.37	Vinnytska:(5); Dnipropetrovska:(2); Donetska:(2); Zakarpatska:(1); Zaporizka:(1); Odeska:(2); Poltavska:(3); Ternopilska:(1); Khmelnytska:(1); Chernihivska:(2)	Vinnytska:(6); Volynska:(2); Dnipropetrovska:(9); Donetska:(3); Zhytomyrska:(5); Zakarpatska:(1); Zaporizka:(8); Kirovohradska:(1); Luhanska:(1); Lvivska:(2); Mykolayivska:(9); Odeska:(3); Poltavska:(8); Rivnenska:(2); Sumska:(1); Kharkivska:(1); Khersonska:(2); Khmelnytska:(4); Cherkaska:(1); Chernihivska:(6)	Vinnytska: (15); Volynska: (9); Dnipropetrovska: (11); Donetska: (5) Zhytomyrska: (7); Zakarpat·ka: (2); Zaporizka: (10); Kyyivska: (4); Kirovohradska: (5); Luhanska: (2); Lvivska: (9); Mykolayivska: (12); Odeska: (14); Poltavska: (25); Rivnenska: (3); Sumska: (12); Ternopilska: (1); Kharkivska: (9); Khersonska: (6); Khmelnytska: (8); Cherkaska:(8); Chernihivska:(10)

Satisfactory level	0.38–0.63	Dnipropetrovska:(3); Poltavska: (2)	Dnipropetrovska:(3); Donetska:(1); Poltavska:(2); Ternopilska:(1); Khmelnytska:(1)	Volynska: (1); Dnipropetrovska (2):; Donetska:(2); Zhytomyrska: (2); Zaporizka: (1); Iv.-Frankivska:(1); Kirovohradska:(1); Lvivska:(1); Mykolayivska:(1); Odeska:(1); Poltavska:(1); Ternopilska:(1); Kharkivska:(1); Khmelnytska:(1); Cherkaska:(1)
Moderate (above average) level	0.64–0.80	Poltavska: (1)	Dnipropetrovska: (1)	Dnipropetrovska: (3) Poltavska:(1)
High level	0.81–1.00	–	–	Dnipropetrovska: (1)

Source: compiled by authors.

UTC of the Dnipropetrovsk region moved from a low level in 2017 to a moderate level in 2018.

Dnipropetrovsk and Poltava regions appeared to be the leaders in the presence of UTC, showing a high and moderate level of financial capacity.

As for such regions as Zakarpattia, Kyivska, Lugansk, Rivne, Sumy, Kherson, Chernivtsi and Chernihiv, all their UTC were at a crisis or low level of financial capacity in 2018. The reasons that most of the united territorial communities of Ukraine are at crisis and low levels can be an intermediate stage of the reform, the reluctance of capable communities to integrate with poor and imperfect systems of distributing the financial support for the reforms.

10.5 Conclusions and Discussion

The reform of decentralization aimed at solving many problems existing at the local level due to the excessive centralization of power and its failure to meet the local needs of society and appropriately improve the quality of public services. Its first stage was focused on creating capable and effective administrative-territorial entities—united territorial communities; its consequences were analyzed in this article.

The estimation of the financial capacity of the united territorial community is an important element of state regulation and management, which takes place both at the local and at the state level. Such results should be considered by the local authorities in defining the main priorities of their activities and in reducing possible financial inequality. The method proposed in this paper allows us to estimate the integral financial state of the united territorial communities and determine their level of financial capacity as an indicator of their financial inequality by using the taxonomic method and clustering the results according to the modified Harrington scale. This made it possible to identify five levels of financial capacity: crisis, low, satisfactory, moderate (above average), and high, which included all the UTC in Ukraine. An analysis of this distribution reveals the tendency that the largest part of the territorial communities fell into a group with a crisis level of financial capacity. It can be applied primarily to the newly formed UTC, which can gradually improve the results and move to another group. The proposed approach concluded that only 3% of the existing united territorial communities have a satisfactory level of financial capacity in 2018.

Acknowledgements and Funding

The research was supported by the Ministry of Education and Science of Ukraine and performed the results of the project "Development of scientific and methodological foundations and practical tools of the financial policy of sustainable development of united territorial communities" (registration number 0117U003935)

References

1. Castells, A. (2001). Document de treball 2000/2: The role of intergovernmental finance in achieving diversity and cohesion: the case of Spain. Environment and Planning C: Government and Policy, 19(2), 189-206. doi.org/10.1068/c0053.
2. Chygryn, O., Petrushenko, Y., Vysochyna, A., Vorontsova, A. (2018). Assessment of fiscal decentralization influence on social and economic development. Montenegrin journal of economics, 14, 4 (2018), 069–084.
3. Decree of the President of Ukraine «About the Strategy of Sustainable Development «Ukraine-2020». (n.d.). zakon.rada.gov.ua. Retrieved from https://zakon.rada.gov.ua/laws/show/5/2015.
4. Goel, R., Mazhar, U., Nelson, M., Ram, R. (2017). Different forms of decentralization and their impact on government performance: Micro-level evidence from 113 countries. Economic Modelling. 62. 10.1016/j.econmod.2016.12.010.
5. Harrington, E. C. (1965). The Desirability Function. Industrial Quality Control, 21, 10, 494–498.
6. Hellwig, Z. (1972). The selection of a set of "core" indicators of socio-economic development. UNESCO.
7. Kanbur, R. and Zhang, X. (2001). Fifty Years of Regional Inequality in China: A Journey Through Revolution, Reform and Openness. CEPR Discussion Paper, 2887. Available at SSRN: https://ssrn.com/abstract=278564.
8. Kasich, A.O., & Petrushko, A.S. (2017). Management of decentralization processes: foreign experience and strategic tasks for Ukraine. Economics and Society, 13, 71–77.
9. Melnyk, L., Sineviciene, L., Lyulyov, O., Pimonenko, T., Dehtyarova, I. (2018), Fiscal decentralization and macroeconomic stability: The experience of Ukraine's economy, Problems and Perspectives in Management, 16(1), 105–114.
10. Order of the Cabinet of Ministers of Ukraine «About the approval of the Concept of Reform of Local Self-Government and Territorial Organization of Power in Ukraine». (n.d.). zakon.rada.gov.ua. Retrieved from https://zakon.rada.gov.ua/laws/show/333-2014-%D1%80.
11. Plyuta, V. (1980). Comparative analysis in economic research: methods of taxonomy and factor analysis. Moscow: Statistics.

12. Pyankovsky, V. (2018). Foreign experience of decentralization and proposals for Ukraine.Actual problems of jurisprudence, 1, 102–106.
13. Pygul, N.G., & Luta, O.V. (2016). Foreign experience of decentralization reforms. Global and national problems of the economy, 9, 684–688.
14. Resolution of the Cabinet of Ministers of Ukraine «On approval of the State Strategy for Regional Development for the period up to 2020 activity». (n.d.). zakon.rada.gov.ua. Retrieved from https://zakon2.rada.gov.ua/laws/show/385-2014-%D0%BF.
15. Resolution of the Cabinet of Ministers of Ukraine «On Approval of the Methodology for the Formation of Capable Territorial Communities». (n.d.). zakon.rada.gov.ua. Retrieved from https://zakon.rada.gov.ua/laws/show/214-2015-%D0%BF.
16. Sember, S.V., & Chubar, O.G., Mashiko, K.S. (2015). Theoretical approaches to the definition of the essence of financial capacity and financial resources of the territorial community. Regional Economics, 3 (77), 81–90.
17. Tkachenko, Yu.V., & Kovalchuk, Ya.O. (2018). An assessment of the effectiveness of decentralization reform: interim results. Collection of scientific works of the ChTTU. Series: Economic Sciences, 48, 21-29.
18. The Law of Ukraine «On Voluntary Association of Territorial Communities» (n.d.). zakon.rada.gov.ua. Retrieved from https://zakon.rada.gov.ua/laws/show/157-19.
19. Tulai, O., Petrushenko, Y., Glova, J., Sydor, I., Ponomarenko, O. (2019) The impact of decentralization on the financial support of regional development. Research Journal "Investment Management and Financial Innovations", 16 (4), 1-15. dx.doi.org/10.21511/imfi.16(4).2019.01
20. Udovychenko, V., Melnychuk, A., Gnatiuk, O., Ostapenko, P. (2017). Decentralization Reform in Ukraine: Assessment of the Chosen Transformation Model. European Spatial Research and Policy. 24. 10.1515/esrp-2017-0002.
21. Vasylieva, T., Harust, Yu., Vynnychenko, N., & Vysochyna, A. (2018). Optimization of the financial decentralization level as an instrument for the country's innovative economic development regulation. Marketing and Management of Innovations, 4, 382-391. http://doi.org/10.21272/mmi.2018. 4–33.
22. Zelinsky, S. E. (2016). Assessment of the capacity of communities in Ukraine. Scientific notes of the Institute of Legislation of the Verkhovna Rada of Ukraine, 6, 128–134.

11

Child Labor Use in Forest Resources Collection and Its Implications on Fertility Preferences in Nepal[1]

Prem B. Bhandari and Medani P. Bhandari

ABSTRACT

Child labor has been utilized to perform various economic activities throughout the human civilization. Basically, poverty is the main cause of child labor; however, in some cases it can be just a traditional family practice. There are varieties of works in which children are involved where they should not be; however, this chapter primarily unveils the social reality of child labor use in forest resources collection and its implication on population growth. Using data from Nepal, we provide evidence that child labor in environmental resources collection contributes to a larger family size desire among individuals. Inequality of gender *per se* is not an independent predictor of individual fertility attitudes. The gender disaggregated analysis suggests that children's participation differentially shapes fertility attitudes among men and women depending upon the types of activities in which children participate. While children's contribution to firewood collection is important among men, their participation in fodder collection is important among women in shaping larger family size. The findings suggest a need for a gender and activity disaggregated focus to address high fertility problems in developing countries like Nepal.

[1] Authors would like to thank Dr. S. S. Shrestha and Dr. Nalini Chhetri for their contributions to the earlier version of this paper that was presented at 2004 PAA annual meeting at Boston. The authors also thank Dr. William G. Axinn, Professor of Sociology, University of Michigan for the data.

11.1 Introduction

> The term "child labor" is often defined as work that deprives children of their childhood, their potential and their dignity, and that is harmful to their physical and mental development. It refers to work that: is mentally, physically, socially or morally dangerous, and harmful to children; and/or interferes with their schooling by: depriving them of the opportunity to attend school; obliging them to leave school prematurely; or requiring them to attempt to combine school attendance with excessively long and heavy work (ILO 2021—https://www.ilo.org/ipec/facts/lang--en/index.htm).

Child Labor "worst forms" under Article 3(d) of C182 are often labeled as "Hazardous forms of child labor" are: (a) all forms of slavery or practices similar to slavery, such as the sale and trafficking of children, debt bondage and serfdom, and forced or compulsory labor, including forced or compulsory recruitment of children for use in armed conflict; (b) the use, procuring or offering of a child for prostitution, for the production of pornography or for pornographic performances; (c) the use, procuring or offering of a child for illicit activities, in particular for the production and trafficking of drugs as defined in the relevant international treaties; and (d) work which, by its nature or the circumstances in which it is carried out, is likely to harm the health, safety, or morals of children (WHO 2017:21; also in Edmonds 2007:10).

The advocacy and research about the child labor problem is not a new phenomenon in academia, as well as in developmental issues. More importantly there is growing concern about the child rights. Much of the economic theorists who had/have discussed about the labor issues, they have discussed about the child labor and its consequences on children development. "Child labor has received considerable attention in economics throughout the discipline's history. Early writings tended to focus on child labor solely through the lens of labor demand. Adam Smith emphasized the value of children in labor shortage societies as motives for fertility. Friedrich Engels wrote extensively on the conditions of working children in the early industrial revolution, and to Marx, child labor was created by the industrial revolution. In his view, machines replaced the need for muscle power, allowing children to do the work formerly performed by men. Though his views on labor supply are not transparent, in Das Kapital he seems to assume that parents and capitalists inevitably" (Edmonds 2007:2) … "Modern writings on child labor are largely based on the human capital theory as developed by T.W. Schultz, Gary Becker, and others. In thinking through the determinants of investments in education, Schultz (1960) emphasized the importance of foregone earnings in human capital accumulation. Investors (parents, children) weigh the return on additional education investments against the costs such investments entail which includes the foregone economic contribution of children. Becker (1962) extended this argument to emphasize that non-wage uses of time are apt to be an equally important influence on the opportunity cost of child's time in school. Rosenzweig and Evenson (1977) appear to be the first published study to apply this framework explicitly to analyze child labor in a developing country context" (Edmonds 2007:3).

After 1970, many scholars begin to raise the concern about the child labor (Schultz 1960; Basu and Zarghamee 2005; Becker 1962, 1965; Becker and Lewis 1973; Rosenzweig 1977; Rosenzweig, and Evenson 1977; Lavalette, 1999; Basu 1999; Edmonds 2008; Rosati, and Rossi, 2003; Gunnarsson et al 2006; Schultz, 2004; Hilowitz 2004; Humphries; 2010; World Bank 1996, 1997, 1988, 1999; ILO 1999a, 1999b, 2000, 2001, 2002, 2006a, 2006b – 2017). As the WHO is the main agency dealing labor issue, "The Sustainable Development Goals include a renewed global commitment to ending child labor. Specifically, target 8.7 of the Sustainable Development Goals calls on the global community to: Take immediate and effective measures to eradicate forced labor, end modern slavery and human trafficking and secure the prohibition and elimination of the worst forms of child labor, including recruitment and use of child soldiers, and by 2025 end child labor in all its forms" (WHO 2017:11). However, it does not seem that, the goal will be met. Similarly, UNCEF has been also working very closely with all concerned stakeholders of the world.

The United Nations Convention on the Rights of the Child is an important agreement by countries who have promised to protect children's rights. The Convention explains who children are, all their rights, and the responsibilities of governments. All the rights are connected, they are all equally important and they cannot be taken away from children" (UNICEF 2021—https://www.unicef.org/child-rights-convention/convention-text-childrens-version). Which defines a child is any person under the age of 18; No discrimination—All children have all these rights, no matter who they are, where they live, what language they speak, what their religion is, what they think, what they look like, if they are a boy or girl, if they have a disability, if they are rich or poor, and no matter who their parents or families are or what their parents or families believe or do. No child should be treated unfairly for any reason. There are 42 main points about the child rights to follow by the government, society, and family: best interests of the child; making rights real; family guidance as children develop; life survival and development; name and nationality; identity; keeping families together; contact with parents across countries; protection from kidnapping; respect for children's views; sharing thoughts freely; freedom of thought and religion; setting up or joining groups; protection of privacy; access to information; responsibility of parents; protection from violence; children without families; children who are adopted; refugee children; children with disabilities; health, water, food, environment; review of a child's placement; social and economic help; food, clothing, a safe home; access to education; aims of education; minority culture, language and religion; rest, play, culture, arts; protection from harmful work; protection from harmful drugs; protection from sexual abuse; prevention of sale and trafficking; protection from exploitation; children in detention; protection in war; recovery and reintegration; children who break the law; best law for children applies; everyone must know children's rights, etc. (UNICEF 2021—https://www.unicef.org/child-rights-convention/convention-text-childrens-version).

> Child labor is one of the most devastating consequences of persistent poverty. Child labor is a complex subject. While often harmful, it is not always so, especially where the alternative is deeper poverty for children and their families (World Bank, IBRB 1998).
>
> "Poverty is the greatest single force which creates the flow of children into the workplace ... [Acute need] makes it nearly impossible for households to invest in their children's education [and] the price of education can be very high [as] most 'free' public education is in fact very expensive for a poor family ... Poor households tend to have more children [and large family size] has been statistically shown to be associated not only with higher likelihood that children work, but also lower school attendance and completion" (ILO 1996b).

These texts and statements clearly accept the challenges of child labor; however, the challenge remains a major concern in the contemporary world. The victimization of child due to conflict, poverty, family violence, and refugee status remains mostly unchanged.

The following few figures of OUR WORLD DATA ORG and WHO 2017 presents the condition of child labor in the world context.

The figures presented above from the OUR WORLD DATA ORG and WHO 2017 present the condition of child labor in the world context. Figures are self-explanatory and provide the alarming picture of reality. There are some positive changes; however, they are not satisfactory. Figure 11.1 gives

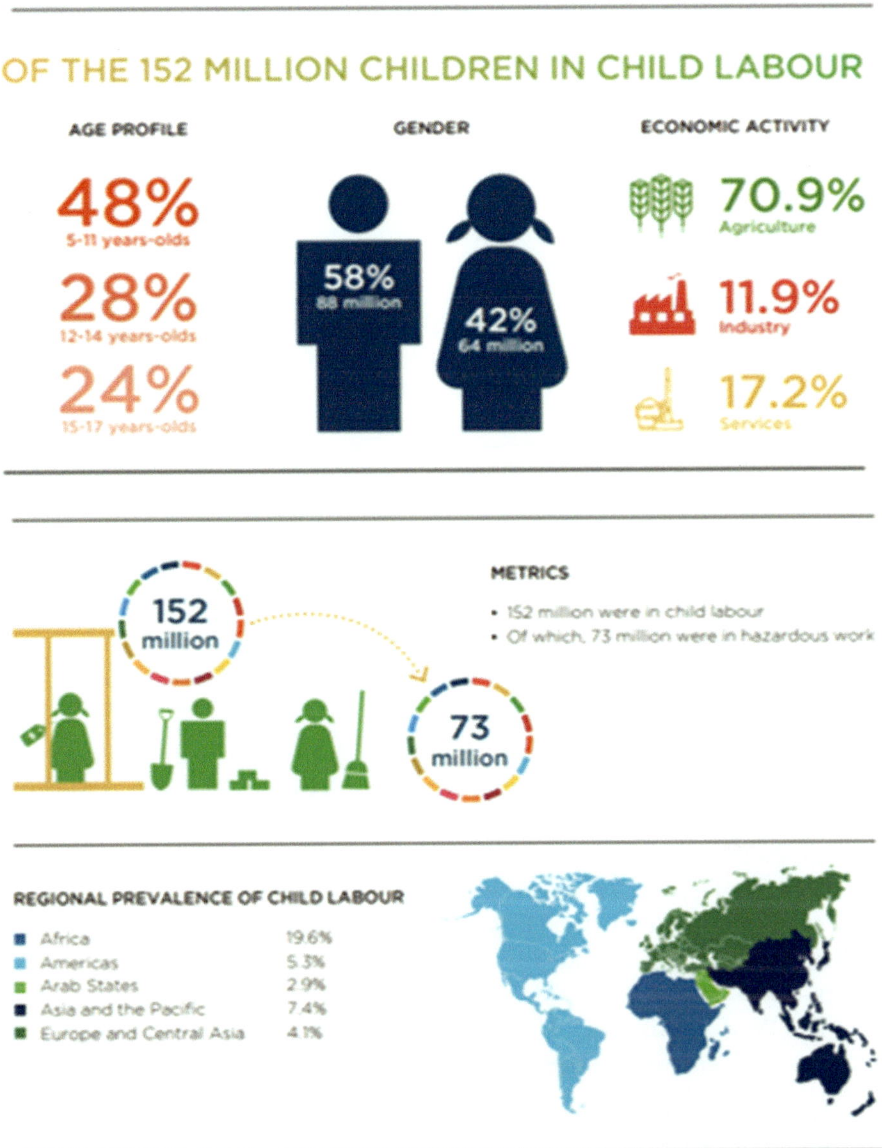

Figure 11.1 Child labor condition—On any given day in 2016 children aged 5–17 years (WHO 2017).
Source: WHO 2017:5

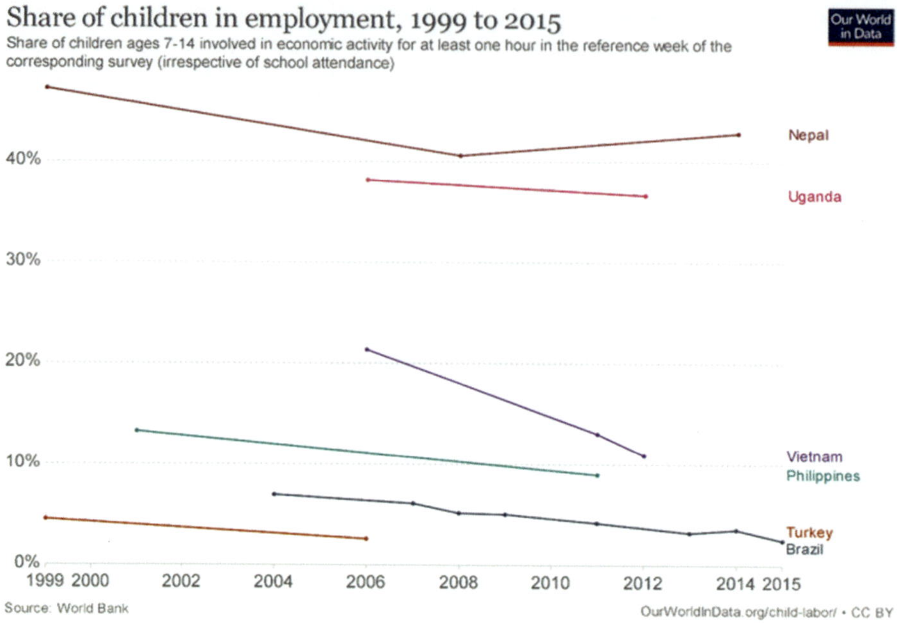

Figure 11.2 Share of children in employment, 1999–2015.
Source: https://ourworldindata.org/child-labor

the overall picture of the scenario (WHO 2017), Figures 11.2, 11.3, and 11.5 show children's schooling and employment, and Figure 11.4 shows the average working hours. A child from Bangladesh has to work about 32 hours for survival. Nepal's children contribute a lot in household income generation, and their involvement is in every sector—farming, small business, industries, and household works. Girls have to do multiple roles, helping their mothers, maintaining the house, and contributing to the labor force of other sectors.

The World Bank, ILO, UNICEF and many scholars have given the way out of how we can reduce the child labor.

Ways to reduce child labor
Reducing poverty
Educating children
Providing support services for working children
Raising public awareness
Legislation and regulation
Promoting elimination through nonregulatory means (World Bank 1998)

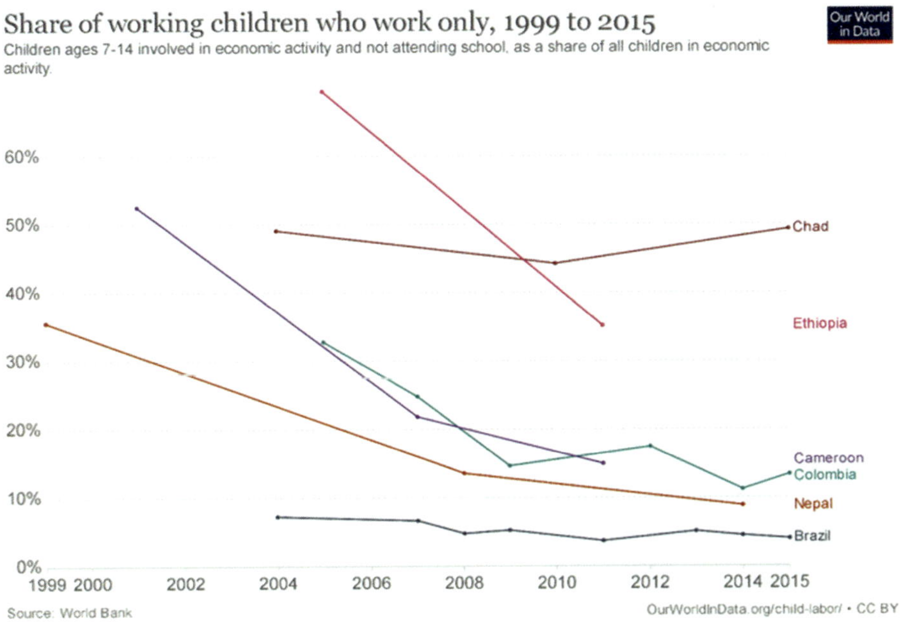

Figure 11.3 Share of working children who work only, 1999–2015.
Source: https://ourworldindata.org/child-labor

Policy responses to child labor need to be integrated into broader national development efforts and adapted to local circumstances.

Policy responses should also address the age, gender, and regional dimensions of child labor.

Continued investment in building the knowledge base on child labor is needed to inform policy responses (WHO 2017).

And most importantly, as noted in the United Nations Convention on the Rights of the Child—provides the clear pathway—how to treat a child. However, documents, research papers, and policies do not solve the problem. To solve the problem, it is necessary to have strong mechanism of policy implementation, and that is still way far. It is the role of every citizen that, the world remains in the shoulders of future generation and today's children are tomorrow's responsible citizens. Therefore, all concerned stakeholders should take responsibility to secure the future by nurturing the today's child.

In this chapter, we address whether children's involvement in forest resources collection contributes to having more children to maintain a household; however, the negative connotation to child growth has been always neglected.

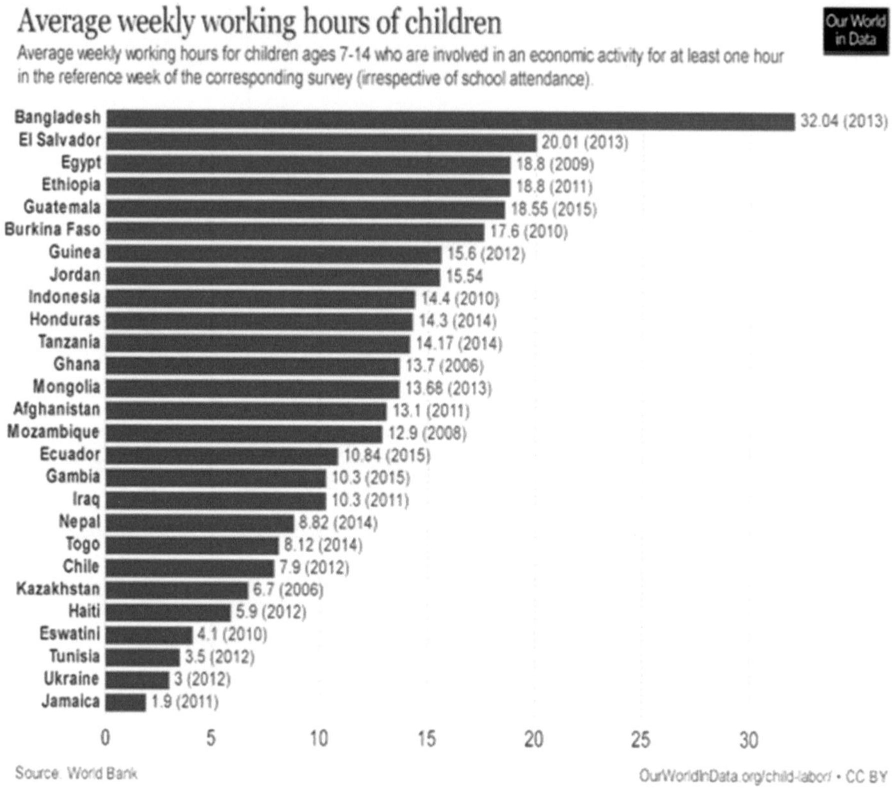

Figure 11.4 Average weekly working hours of children.
Source: https://ourworldindata.org/child-labor

11.2 Nepal Case

Since the 1970s, Nepal has witnessed the onset of fertility transition (Acharya 1998; Sharma and Kayastha 1998; UN 1997). Yet the country remains in the grip of high fertility. Recent statistics shows that the total fertility rate (TFR) remained high at 4.5 in 2003 which has declined to 2.3 in 2016 (Ministry of Health, Nepal; New ERA; and ICF 2017). Addressing the problem of high fertility has been a priority for the Nepalese government, which warrants a better understanding of the factors governing high fertility situation in the country.

Nepal's economy is largely agriculturally based. In an agrarian society where the labor market is not well developed, children's labor contribution to household production activities is substantial. Children are widely used in performing various household activities including the collection of forest

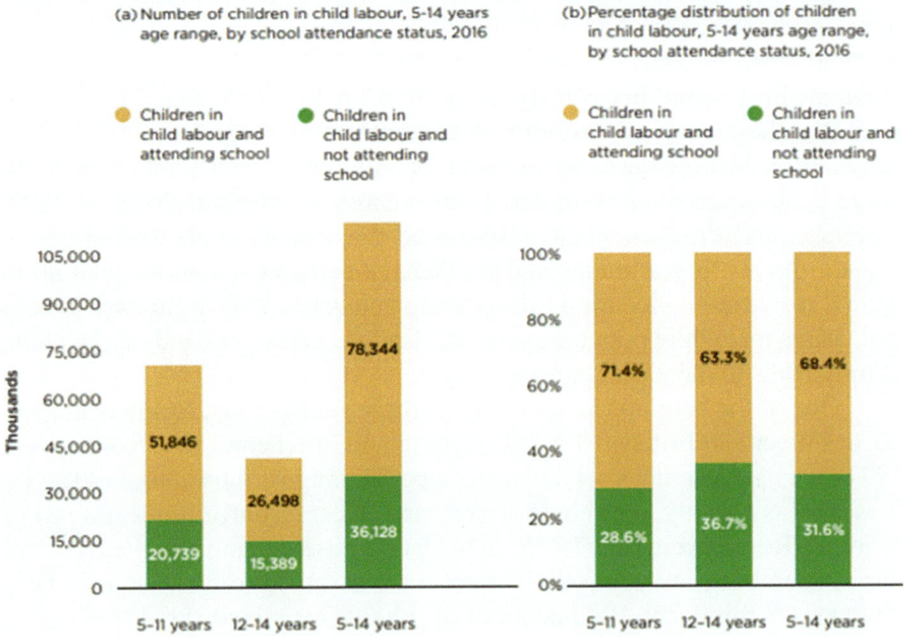

Figure 11.5 Child labor and school attendance.
Source: WHO 2017:47

resources (Bhandari et al. 1996; Chitrakar 1990; Filmer and Pritchett 1997; Karan and Ishii 1996; Kumar and Hotchkiss 1988; Loughran and Pritchette 1997). It is because the household economy is intimately dependent on natural resources such as firewood for energy and fodder for maintaining livestock and farming (Bhandari et al. 1996: Gurung 1987; Yadav 1990). Since the labor demand for the collection of such resources has to be fulfilled to a greater extent from within the household, one of the ways to meet the potential labor demand in a household may be to produce more children. The demand for child labor in a household is one of the explanations for high fertility among individuals of agrarian societies (Filmer and Pritchett 1997; Loughran and Pritchette 1997; Rosenzweig 1977; Rosenzweig and Evenson 1975). Thus, understanding the relationship between child labor use and human fertility in relation to the consumption of natural (environmental) resources is vital.

Although various explanations have been forwarded to explain high fertility in developing countries, the frameworks provided by Rosenzweig (1977) and Stokes and his colleagues (e.g., Stokes and Schutjer 1984; Stokes,

Schutzer, and Bulatao 1986) focus on the demand for children in agricultural households. While Rosenzweig (1977) considers psychic and productive roles of children as important factors in influencing the demand for children, Stokes and his colleagues in their land–labor demand hypothesis posit that the greater accessibility to cultivated land requires more laborers, which ultimately increases the demand for children. Recently, other researchers (e.g., Biddlecom, Axinn, and Barber 2000; Filmer and Pritchett 1997; Loughran and Pritchette 1997) have used the vicious circle framework to explain the fertility attitudes and the behavior of individuals in relation to natural resources collection in developing countries. This argument, in general, states that while population growth leads to environmental degradation, in turn, leads to increased human fertility.

One of the common conclusions of these studies (e.g., Biddlecom et al. 2000; Filmer and Pritchett 1997; Loughran and Pritchette 1997; Rosenzweig 1977) is that children are widely used in performing various household activities such as farming, household chores, and forest resources collections. For example, Rosenzweig (1977:125) in his farm household model notes that "all family members, including the children, are assumed to contribute to farm output … " Similarly, Biddlecom et al. (2000) examined the difference in fertility attitudes of individuals between those who purchased firewood and those who collected it with the assumption that all the households that collected firewood use child labor. However, these studies have overlooked the fact that not all households use child labor in performing certain household activities such as farming and natural resources collection. Hence, the extent to which child labor use in performing various activities affects fertility attitudes among individuals of environmental resources collecting households is specifically unexplained. Moreover, these studies have mainly focused on women (except Biddlecom et al. 2000).

Considering this background, we attempted to answer two main questions: (1) Does the use of child labor in environmental resources collection, particularly *firewood* and *fodder*, contribute to individual fertility preferences? (2) Does the use of child labor in collecting these resources shape different fertility preferences among men and women? We hypothesized that individuals in a household that uses child labor prefer larger family sizes than those who do not. Moreover, the extent to which child labor substitutes for either parent's time in different activities can be associated with the difference in individual's fertility preferences. Since children are assumed to substitute particularly for women's work, we further hypothesized that women prefer larger family sizes than men.

11.3 The Framework

Various explanations such as the household demand framework (Becker 1965, 1981; Rosenzweig 1977), the socioeconomic framework (Easterlin 1975; Easterlin and Crimminis 1985), the land–labor demand hypothesis and land-security hypothesis (Stokes and Schutjer 1984; Stokes, Schutzer, and Bulatao 1986), the wealth flows theory (Caldwell 1982; Caldwell and Caldwell 1987), and few others have been used in explaining the fertility behavior of individuals. These frameworks directly or indirectly focus on the economic and non-economic contribution of children to households resulting in a greater demand for children.

We base this study on the household demand framework of Becker (1965; 1981), Rosenzweig and Evenson (1975), and Rosenzweig (1977) extending the role of child labor use in household activities and its impact on individual fertility preferences. Becker (1965; 1981) argued that children's contribution to family income by engaging in household and market activities increases benefits and hence, reduces the cost of raising them. Given household income constraints, a decrease in costs of children leads to a desire for more children. Conversely, it can be argued that increased benefits from children in a household may increase the demand for them. This is a much broader framework that focuses on the contribution of children to a household and its effects on the fertility behavior of couples.

Rosenzweig (1977) and Rosenzweig and Evenson (1975) emphasized the importance of children in agrarian households and high human fertility among individuals resulting from the greater demand for them. Rosenzweig (1977:124) pointed out two important roles of children in an agrarian society—"as durable commodities which yield psychic income and productive laborers … ." Emphasizing the roles of children in performing various economic activities in household production, he argued that Becker's model fails to incorporate a child's role in household production models. Using the aggregate U.S. farm population cross-sectional data from 1939–1960, he observed that the decline in birth rates was associated with the reduction of value of children in farm households as a result of the application of capital-biased technical change. The use of technologies decreased the value of child labor and hence reduced the pecuniary returns from children within agriculture sector. This resulted in the decline in farm birth rates. He further pointed out that increases in non-farm employment opportunities decreased the demand for farm children due to the loss of the valuable services of mature farm children by the farm family contributing to a decrease in birth rates. Although these results are based on the study carried out by using the U.S. based data, the farm fertility model presented by Rosenzweig may be applied to agricultural societies in developing countries such as Nepal.

In an agrarian society of many developing countries, households are both a production and a consumption unit. Because of the absence of labor markets for household production activities, some households strategically use child labor to perform daily reproduction or maintenance activities, including the collection of environmental resources such as firewood for energy and fodder for livestock. Such a strategy not only contributes to the production of household goods and services, but also substitutes for time of other members of a household engaged in the household production process. Substitution of adults' time allows them to diversify their labor allocation to other productive activities such as off-farm employment. For such households, production of children may be beneficial. Then, there are other households that do not use child labor in household production process. For these households, investing in children's human capital development such as education can be an alternative strategy with the anticipation of obtaining long-term benefits from them. Such a strategy may increase the cost of raising children and therefore, may have different effects on fertility decisions. For instance, since an investment in children's education is costly, individuals belonging to these households may prefer to have fewer children. Given this background, *we argue that individuals who live in households that use children in household production processes, particularly for collecting firewood and fodder, prefer larger family sizes than those who do not use them for such purposes.*

It has been realized that previous fertility studies lack in appreciating the roles of both men and women in fertility decisions (Biddlecom et al. 2000; Greene and Biddlecom 2000) and therefore focus on women only. Thus, the next issue we are concerned with is whether there are any apparent gender differences in fertility attitudes of individuals. Mason and Taj (1987) reported that there may be differences in fertility goals by gender under certain conditions. One of the important conditions is the women's and men's social and/or economic relationship to children. For example, it is generally assumed that children's participation in forest resources collection to a greater extent substitutes for women's time. Given this assumption, *we further argue that women prefer larger family sizes than men.*

11.4 The Setting

The Western Chitwan Valley situated in the southern plain of central Nepal is the setting for this study. Before the 1950s, the Valley was mainly inhabited by indigenous Tharu ethnic group. It was covered with dense forests and was infamous for malarial infestation. The government, with the assistance from the USA, initiated a rehabilitation program in the Valley during the 1950s by clearing the dense forests. Since then, the Valley has witnessed a rapid inflow of migrants of diverse ethnic groups. Further, Chitwan's central location and relatively well-developed transportation network have been the catalytic forces for turning it into a hub for business and tourism. This has resulted in a rapid proliferation of government services, businesses, and wage labor jobs in the district (Shivakoti et al. 1999).

Farming is the main source of livelihood of people in the Valley. Although agriculture is experiencing rapid modernization, it is largely subsistence in nature. A large majority of farmers practice mixed farming with highly integrated crop and livestock production systems. A household survey conducted in the valley in 1996 indicated that of the total 1,805 sample households, over 80% of them were growing some crops and about three-fourths of them were keeping livestock such as cattle, buffalo, sheep, and goats.

The forest is an integral part of the farming household. Environmental resources such as firewood, fodder, and timber directly come from forests. Extraction of these resources occurs mainly from the Royal Chitwan National Park and the adjoining forests of Barandabar, Kalaban, and community forests. According to the same 1996 household survey, about 92% of the households were currently using firewood for energy. Among them, a large majority of these households (about 82%) reported that they partly or wholly collected firewood from the public and private forests and common lands. Similarly, almost all of the livestock keeping households (98%) collected all or a part of the fodder required for their animals from forest and common lands.

Collection of firewood and fodder demands a substantial amount of time for the households. The 1996 survey showed that among the firewood-collecting households, women collected firewood from 86% of them. The percentages of households with men's and children's participation in firewood collection were 70 and 12, respectively. For fodder collection, over 95% of the households used female labor as compared to the 55% of the households that used male labor. About one-thirds of the households used child labor.

11.5 Data

The data for this study is drawn from the Chitwan Valley Family Study (CVFS) and the Population and Environment Study conducted in the Western Chitwan Valley of Nepal by the Population and Ecology Research Laboratory (PERL) in 1996. Barber et al. (1997) have described the sampling procedure used for these studies in detail. In brief, a total of 1,805 households were surveyed. At the individual level, information was collected from a total of 5,271 persons between 15 and 59 years of age and their spouse living in the sample households. In this study, we used information from—(i) all currently married individuals, (ii) between 15 and 45 years, (iii) with at least one living child, and (iv) living in a household that collected firewood and/or fodder during the survey year. Within this specification, there were a total of 2,008 individuals who belonged to the households that collected firewood and 2,035 individuals who belonged to the households that kept livestock and collected fodder. In both of these samples, 57% of them were females. Of the total, about 13% and 35% of the individuals respectively belonged to the households that used children for firewood and fodder collection activities.

11.6 Variable Measurement and Methods of Analysis

In this study, the dependent variable is the Coombs-scaled measure of fertility preference (Coombs 1974) that approximates an ordinal-level measure of fertility preference. It is considered as one of the most accurate predictors of subsequent fertility behavior as compared to the commonly used measure of ideal family size (Coombs 1974; Farooq 1985; Thompson and Brabdreth 1995). The measure of an ideal family size does not provide variation in the intensity of desire among those who want the same number of children.

The Coombs scale measure of fertility preference was measured by asking a series of questions. The first question was, *"People often do not have exactly the number of children they want to have. If you could have exactly the number of children you want, how many children would you want to have?"* After getting the response to this question, a subsequent question was asked, *"If you could not have ----- (number of children from the above question) children, would you want to have ----- (one number lower than the first response) or ------- (one number higher than the first response)?"* Based on the response of the second question, the third question was *"If you could not have ----- (number of children from the second question) children, would you want to have ----- (one number lower than the second response) or ------- (one number higher than the second response)?"* The responses thus recorded were converted to the Coombs scale ranging from 1 to 25. The value 1 represents the lower boundary and the value 25 represents the upper boundary of an individual's fertility preference.

We considered two major independent variables to examine their effects on individual fertility attitudes: child participation in environmental resources collection and gender. The first independent variable is whether or not a household used child labor for the collection of firewood and/or fodder. This variable is dichotomously coded as "1" if a household used child labor in collecting firewood and/or fodder and "0" if otherwise. To examine the independent effect of child labor use on fertility preferences, we also controlled for the effects of the participation by adult men and women in firewood and fodder collection in a household. The participation by men and women is also dichotomously coded as "1" if a household used male/female labor in collecting firewood and/or fodder and "0" if otherwise.

We expected that the number of children an individual desires is governed by what an individual perceives about the potential benefits and costs of having children. Because of differential gender roles and responsibilities in a household, the services that parents obtain from their children can be different when we compare a mother and a father. The potential costs and benefits of children that a mother perceives can be different from what a father does. Therefore, we also examined the effect of gender on family size preferences. This variable is dichotomously coded as "1" for female and "0" for male.

We controlled other individual and household level factors to net out the effects of the use of child labor and the gender on fertility preferences. The mechanism of how these variables affect individual fertility goals and the specification of these variables is briefly described.

Individual's age, one of the proximate determinants of fertility (Bongaarts 1978; Davis and Blake 1956), is closely associated with activities such as marriage, divorce, and frequency of intercourse that affect fertility behavior of an individual (Simmons 1985). To a married couple or an individual, an increase in age increases the likelihood of an individual in sexual union. Therefore, a positive effect of age on fertility behavior is expected. We also controlled the effect of the number of living children.

Child mortality is another important issue associated with high fertility in developing countries. If an individual has experienced the death of a child in his/her lifetime, it is expected that his/her fertility preference should be greater than those who have not experienced any. It is because individuals who have experienced child mortality may wish to replace a child or increase the number of children as future insurance (Preston 1978). In this study, child mortality experienced by an individual is measured as "1" if any child had died in her/his life and "0" if otherwise.

Effect of education on fertility is widely studied and these are found to be negatively associated (Becker 1981; Easterlin and Crimminis 1985). Similar findings have been reported by Pearce (2000) and Gajurel (2001) in this study setting. An individual's education is measured as the number of years of schooling at the time of survey. Employment of an individual also influences individual fertility behavior. For example, an increase in non-farm employment opportunities decreases the demand for farm children due to the loss of the valuable services of children when they mature (Rosenzweig 1977). Moreover, the opportunity cost of time of employed individuals with salaried job is greater than those of the unemployed. For example, those who are in salaried jobs, children can be costlier. Labor force participation especially for women is considered as an indicator of status of autonomy that negatively affects fertility (Mason 1987; Waite and Stolzenberg 1976). In this study, employment status of an individual is measured as whether or not an individual is currently employed in a salaried job.

Increased child raising costs reduce the demand for children (Becker 1981; Robinson 1997). In this study, we used the perception of an individual about the cost of raising a child. In the survey, it was measured by asking, *"Is raising children cheap these days?"* Responses were recorded as highly agree, agree, disagree, and highly disagree. For the purpose of analysis, we recoded this variable as "1" referring to raising children is expensive and "0" if otherwise. Exposure to mass media such as listening to family planning messages on radio and/or television (TV) is another factor that influences the reproductive behavior of individuals. Exposure to communication media increases the exposure to modern ideas about family, childbearing, consumption, and individualism that are communicated through these channels (Westoff and Bankole 1999). Researchers suggest that mass media are important sources of information about contraception in developing countries, whereby individuals improve their knowledge and shape perceptions and attitudes about various subjects that influence sexual and reproductive behavior (Thapa and Mishra 2003). This variable is measured dichotomously as "1" if ever listened to family planning programming on radio and "0" if never listened to such programs.

In a farming society, organization of agriculture such as land ownership and access to cultivated land influence human fertility (Thomas 1991; Stokes et al. 1986; Stokes and Schutjer 1984). The land–labor demand hypothesis suggests that households with relatively greater access to land require more labor to cultivate it. Therefore, individuals living in a household with a large amount of cultivated land demand more children. Tuladhar et al. (1982) and Gajurel (2001) provide evidence of the land–labor demand hypothesis from Nepal. The access to cultivated land is measured in *kattha*, a local unit (30 *kattha* = 1 hectare). The next control variable we considered was the availability of toilet facilities in the home. This variable is used as an indicator of socioeconomic status of a household (Montgomery et al. 2000; Pebley, Goldman, and Rodriguez 1996; UN 1997). In the Chitwan Valley, generally, relatively well-off households tend to have toilet facilities in the home. Considering that individuals of higher socioeconomic status tend to have fewer children, individuals living in a household with toilet facilities are expected to desire fewer children than those without such facilities. This variable is measured as "1" if toilet facilities are available in a household and "0" if not.

We also took into account the effects of technology use, migration and children's schooling on fertility preferences as used by Rosenzweig (1977). He indicated that the use of labor-saving technologies changes parameters of production by influencing children's productivity. Therefore, uses of labor-saving technologies affect demand for children negatively. Use of labor-saving technology is measured as whether a household used bicycles and/or carts in transporting firewood/fodder or not. On the effect of migration, Rosenzweig (1977) opined that children are net assets to households when they become mature. However, if the likelihood that children will migrate later is high, then they are no longer considered to be assets and households will consequently have less demand for them. Migration of individuals from a household is measured as to whether any individual is currently away from home for work reasons or not. Rosenzweig (1977) further pointed out that children's schooling also reduces demand for them. Since we do not have a measure of whether a child has participated in school, we used time to walk (in minutes) to the nearest school. It is because the availability and proximity of schools also encourages individuals toward fertility limitation (Axinn and Barber 2001). In their view, households located in closer proximity to schools are more likely to send their children for schooling, which, on the one hand, reduces the contribution of children to household activities, and on the other increases the cost of raising a child. Therefore, we expected that the greater distance to school is associated with larger family size preferences.

The Chitwan Valley is resided by people of various ethnic groups such as High Caste Hindu (e.g., Brahmin and Chhetri), Low Caste Hindu (e.g., Kami, Damai, Sharki), Newar, Hill Tibetoburmese (e.g., Gurung and Magar), and indigenous Terai Tibetoburmese groups (e.g., Tharu, Kumal, and Darai). Previous studies in this research setting observed a variation in fertility behaviors of individuals by ethnicity (Axinn and Barber 2001; Biddlecom et al. 2000; Gajurel 2001; Pearce 2000). Individuals belonging to the local indigenous or Terai Tibetoburmese ethnic groups preferred significantly larger family sizes compared to other ethnic groups. Therefore, we also adjusted for the effect of ethnicity in this study. We categorized ethnicity into two broad groups: individuals belonging to the indigenous or Terai Tibetoburmese groups and Others (High Caste Hindu, Low Caste Hindu, Newar, and Hill Tibetoburmese).

11.7 Analytic Strategy

As the dependent variable Coombs-scaled fertility preference is continuously measured, we used the ordinary least square (OLS) technique as a multivariate tool in order to tease out the net effects of child labor participation and gender adjusting for other factors. First, we performed a pooled analysis for individuals (including both men and women) who belonged to the households that collected two environmental resources, *firewood* and *fodder,* separately. In both of these activity-level pooled analyses, we ran five different models considering different sets of variables to examine the consistency of the effects of independent variables on fertility preference. In model 1, we examined the effects of child labor use and gender on fertility preferences. In models 2 through 4, we examined the net effects of these two major variables, controlling for the effects of individual characteristics (except education and employment), household characteristics, and uses of adult men and women labor, respectively. In model 5, we further added two individual level variables—education and employment status of individuals.

Second, we disaggregated the activity-level analysis by gender to examine the effect of child labor use on fertility preferences among men and among women. We ran three different models in each men and women group. In each group, in the first model, we examined the effect of child labor use on fertility preference. In other two models, we controlled the effects of individual and household-level characteristics. Next, in order to ascertain the effect of gender, additional models were run among users and non-users of child labor in environmental resources collections. Since the effects of gender in user and no-user models are not statistically significant, we have not reported these results in the paper to save space.

11.8 Results and Discussion

11.8.1 Descriptive Results

The descriptive statistics of the variables used in the analysis of firewood and fodder collection activities, broken down by child labor use and gender, are provided in Tables 11.1 and 11.2, respectively. In firewood collection, the mean preference of individuals for children (7.61) in Coombs's scale is significantly greater for those from households using child labor as compared to

6.65 for those from households that did not use child labor (one-way ANOVA $F = 33.069$; $p<.001$; Table 11.1). The results were similar among fodder collectors (Table 11.2). Analysis by gender showed that for both firewood and fodder collecting activities, women preferred slightly larger family sizes as compared to men. For example, among firewood collectors, the average preference in the Coombs scale was 6.93 for women and 6.56 for men (one-way ANOVA $F = 10.677$; $p<.001$). Among fodder collectors, the average preference was 6.80 for women and 6.52 for men (one-way ANOVA $F = 6.140$; $p<.05$).

With regard to other variables, in both firewood and fodder collection activities, on an average, individuals living in households that used child labor were relatively older, had larger number of living children, were less educated, and a relatively higher proportion of them had experienced child mortality as compared to those who did not use child labor. In addition, they were proportionately less exposed to family planning programming on radio. From this category, fewer individuals had toilet facilities in the home, and were likely to be from households with a large size of cultivated landholding, and a smaller proportion of them had used mechanical means (bicycle and cart) in transporting firewood and fodder. In general, women were relatively younger with a lower average number of schooling, and a relatively smaller proportion of them were exposed to the family planning programming on radio than were men.

11.8.2 Multivariate Results

11.8.2.1 Child Labor Use and Fertility Preferences

Tables 11.3 and 11.4 provide unstandardized ordinary least square regression estimates for predicting the effect of child labor use in firewood and fodder collection, respectively, on fertility preferences of currently married individuals of 15–45 years of ages with at least one living child controlling for the effects of other factors. For both firewood and fodder collection activities, the use of child labor positively and significantly contributed to larger family size desires among individuals who lived in households that used child labor (model 1, Tables 11.3 and 11.4). In firewood collection, individuals from households that used children had 0.96 ($p<.001$) points higher level of preference in the Coombs scale over those from households that did not use child labor net of gender effect. Similarly, in fodder collection, adjusting for the effect of gender, the use of child labor contributed to over 0.88 ($p<.001$) points higher level of preference for those who used children compared to those who did not.

Table 11.1 Descriptive statistics of variables used in models for firewood collection

	Child Labor				**Gender**			
	Used (n = 254)		**Not Used (n = 1754)**		**Women (n = 1146)**		**Men (n = 862)**	
Variables	**Mean**	**SD**	**Mean**	**SD**	**Mean**	**SD**	**Mean**	**SD**
Dependent variable								
Fertility preference (Coombs scale)	7.61	2.88	6.65	2.43	6.93	2.55	6.56	2.46
Independent variables								
Child labor (used = 1)	-	-	-	-	0.13	0.33	0.13	0.33
Gender (female = 1)	0.57	0.50	0.57	0.50	-	-	-	-
Individual characteristics								
Age (years)	35.17	6.67	31.52	7.24	31.16	7.64	33.08	6.60
Number of living children	3.71	1.70	2.93	1.50	3.09	1.59	2.95	1.50
Child ever died (yes = 1)	0.32	0.47	0.22	0.41	0.25	0.43	0.21	0.41
Education (years of schooling)	2.57	3.71	3.87	4.35	2.24	3.50	5.66	4.48
Currently on salaried job (yes = 1)	0.15	0.35	0.14	0.35	0.00	0.09	0.32	0.47
Perceived child raising cost (expensive = 1)	0.67	0.47	0.68	0.46	0.61	0.49	0.77	0.42
Exposed to family planning programming on radio (yes = 1)	0.73	0.44	0.80	0.40	0.75	0.43	0.84	0.37
Household characteristics								
Cultivated land (kattha)	29.72	29.48	25.39	24.97	26.35	25.79	25.39	25.39
Toilet facilities (yes = 1)	0.54	0.50	0.64	0.48	0.63	0.48	0.62	0.49
Migration for work (yes = 1)	0.26	0.44	0.28	0.45	0.29	0.45	0.27	0.45
Distance to school (minutes)	9.34	6.66	9.99	8.01	9.92	7.95	9.90	7.74
Collection of firewood (non-foot = 1)	0.06	0.24	0.12	0.32	0.11	0.32	0.11	0.32
Alternative labor use								
Female labor (used = 1)	0.91	0.29	0.87	0.34	0.88	0.33	0.87	0.33
Male labor (used = 1)	0.74	0.44	0.74	0.44	0.72	0.45	0.75	0.43
Ethnicity (reference = others)								
Terai Tibetoburmese	0.37	0.48	0.20	0.40	0.21	0.41	0.24	0.43

Note: 30 kattha = 1 hectare

Table 11.2 Descriptive statistics of variables used in models for fodder collection

	Child Labor				Gender			
	Used (n = 704)		Not Used (n = 1331)		Women (n = 1170)		Men (n = 865)	
Variables	Mean	SD	Mean	SD	Mean	SD	Mean	SD
Dependent variable								
Fertility preference (Coombs scale)	7.26	2.70	6.37	2.30	6.80	2.39	6.52	2.60
Independent variables								
Child labor (used = 1)	-	-	-	-	0.36	0.48	0.33	0.47
Gender (female = 1)	0.59	0.49	0.57	0.50	-	-	-	-
Individual characteristics								
Age (years)	35.02	6.56	30.09	7.26	31.57	7.69	33.36	6.59
Number of living children	3.74	1.62	2.67	1.35	3.11	1.58	2.95	1.47
Child ever died (yes = 1)	0.27	0.45	0.20	0.40	0.24	0.43	0.21	0.41
Education (years of schooling)	3.09	4.07	4.48	4.54	2.38	3.58	6.20	4.53
Currently on salaried job (yes = 1)	0.14	0.35	0.15	0.35	0.01	0.09	0.33	0.47
Perceived child raising cost (expensive = 1)	0.69	0.46	0.69	0.46	0.62	0.49	0.78	0.41
Exposed to family planning programming on radio (yes = 1)	0.78	0.41	0.82	0.38	0.78	0.42	0.85	0.35
Household characteristics								
Cultivated land (kattha)	31.42	29.64	26.36	23.28	28.46	25.85	27.64	25.66
Toilet facilities (yes = 1)	0.63	0.48	0.69	0.46	0.67	0.47	0.66	0.47
Migration for work (yes = 1)	0.25	0.43	0.32	0.47	0.30	0.46	0.29	0.46
Distance to school (minutes)	9.42	8.42	9.75	7.27	9.60	7.72	9.68	7.64
Collection of fodder (non-foot = 1)	0.05	0.21	0.07	0.26	0.06	0.25	0.06	0.24
Alternative labor use								
Female labor (used = 1)	0.95	0.21	0.95	0.21	0.95	0.21	0.95	0.22
Male labor (used = 1)	0.56	0.50	0.56	0.50	0.56	0.50	0.57	0.50
Ethnicity (reference = others)								
Terai Tibetoburmese	0.23	0.42	0.17	0.38	0.18	0.39	0.20	0.40

Note: 30 kattha = 1 hectare

Table 11.3 Unstandardized OLS regression estimates predicting the effect of child labor use in firewood collection on the fertility preferences of currently married individuals of age 15–45 years with at least one living child (N = 2008)

Variables	**Model 1**	**2**	**3**	**4**	**5**
Independent variables					
Child labor (used = 1)	0.963***	0.503**	0.343*	0.347*	0.329*
Gender (female = 1)	0.369***	0.229*	0.320**	0.338**	0.090
Individual characteristics					
Age (years)	-	0.007	0.025*	0.027**	0.017^{+}
Number of living children	-	0.435***	0.372***	0.372***	0.358***
Child ever died (yes = 1)	-	0.582***	0.467***	0.458***	0.400**
Perceived child raising cost (expensive = 1)	-	-0.206^{+}	-0.214^{+}	-0.201^{+}	-0.152
Exposed to family planning programming on radio (yes = 1)	-	-0.520***	-0.319*	-0.313*	-0.221^{+}
Education (years of schooling)	-	-	-	-	-0.085***
Currently on salaried job (yes = 1)	-	-	-	-	0.167
Household characteristics					
Cultivated land (kattha)					
Toilet facilities (yes = 1)	-	-	0.001	0.003	0.000
Migration for work (yes = 1)	-	-	-0.423***	-0.406***	-0.271*
Distance to school (minutes)	-	-	-0.016	-0.066	0.038
Collection of firewood (non-foot = 1)	-	-	0.002	0.002	-0.001
Ethnicity (reference = others)	-	-	0.146	0.090	0.104
Terai Tibetoburmese					
Alternative labor use	-	-	0.806***	0.757***	0.600***
Female labor (used = 1)	-	-	-	0.048	0.006
Male labor (used = 1)	-	-	-	0.403***	0.350**
Intercept	6.436***	5.453***	4.993***	4.593***	5.308**
Model F	22.038***	42.262***	28.904***	25.846***	24.846***
Regression degrees of freedom	2	7	13	15	17
Residual degrees of freedom	2005	2000	1994	1992	1990
Adjusted R-square	2.1%	12.6%	15.3%	15.7%	16.8%

Note: 30 kattha = 1 hectare

*** = p<.001; ** = p<.01; * = p<.05; + = p<.10

Table 11.4 Unstandardized OLS regression estimates predicting the effect of child labor use in fodder collection on the fertility preferences of currently married individuals of age 15–45 years with at least one living child (N = 2035)

	Model				
Variables	**1**	**2**	**3**	**4**	**5**
Independent variables					
Child labor (used = 1)	0.883***	0.362**	0.296**	0.296**	0.273*
Gender (female = 1)	0.256*	0.105	0.176	0.176	-0.075
Individual characteristics					
Age (years)	-	0.008	0.024	0.023*	0.014
Number of living children	-	0.398***	0.349***	0.354***	0.341***
Child ever died (yes = 1)	-	0.570***	0.474***	0.468***	0.415***
Perceived child raising cost (expensive = 1)	-	-0.330**	-0.339**	-0.330**	-0.276*
Exposed to family planning programming on radio (yes = 1)	-	-0.558***	-0.389**	-0.902**	-0.320*
Education (years of schooling)	-	-	-	-	-0.067***
Currently on salaried job (yes = 1)	-	-	-	-	0.020
Household characteristics					
Cultivated land (kattha)	-	-	-0.001	0.000	0.001
Toilet facilities (yes = 1)	-	-	-0.405***	-0.401***	-0.297*
Migration for work (yes = 1)	-	-	-0.036	-0.001	-0.006
Distance to school (minutes)	-	-	0.002	0.001	-0.002
Collection of fodder (non-foot = 1)	-	-	-0.161	-0.194	-0.137
Ethnicity (reference = others)					
Terai Tibetoburmese	-	-	0.691***	0.692***	0.559***
Alternative labor use					
Female labor (used = 1)	-	-	-	0.363	0.360
Male labor (used = 1)	-	-	-	0.155	0.125
Intercept	6.229***	5.570***	5.260***	4.823***	5.439***
Model F	33.145***	40.460***	26.414***	23.162***	21.756***
Regression degrees of freedom	2	7	13	15	17
Residual degrees of freedom	2032	2027	2021	2019	2017
Adjusted R-square	3.1%	12.0%	14.0%	14.0%	14.8%

Note: 30 kattha = 1 hectare
*** = p<.001; ** = p<.01; * = p<.05; + = p<.10

When the effects of other variables such as individual and household-level characteristics were adjusted (models 2 through 5 of Tables 11.3 and 11.4), despite decrease in both the magnitude of the coefficients and the strength of statistical significance, the effect of child labor use still remained statistically significant in both firewood and fodder collection models. For instance, net of other factors, individuals from households using child labor preferred significantly greater family size desires by 0.33 ($p<.05$) and 0.29 ($p<.05$) points in Coombs scale, respectively, for firewood and fodder collection compared to those from households not using child labor (model 5 of Tables 11.3 and 11.4). The results provide evidence in support of our argument that the use of child labor in forest resources collection contributes to shaping larger family size desires among individuals. This result provides an additional insight in explaining fertility preferences of individuals in developing countries in that the fertility attitudes of individuals differ not only between those who purchase and those who collect environmental resources (e.g.,, Biddlecom et al. 2000) but also among collectors who use or do not use child labor.

11.8.2.2. Gender and Fertility Preferences

We also examined whether fertility preferences of individuals living in environmental resources collecting households vary by gender. Our framework suggests that as firewood and fodder collection activities are the primary responsibilities of women, child labor substitutes mainly for women's time. Therefore, we expected that women should have a significantly larger family size desires compared to men in a household that collected firewood and fodder using child labor.

The results of the effect of gender on individual family size desires are provided in Tables 11.3 and 11.4, respectively, for firewood and fodder collection activities. The results in model 1 (Tables 11.3 and 11.4) showed that women preferred significantly larger family sizes than men. When the effects of all other factors such as individual and household-level characteristics were controlled for (particularly education and employment in firewood collection models), the statistical significance of gender effect disappeared, however (model 5 of Table 11.3). One possible explanation could be that both education and employment might have confounded the effect of gender. The results of descriptive statistics (Tables 11.1 and 11.2) show that men have significantly higher level of schooling and a higher proportion of them are currently employed than women. Further, there was no clear effect of gender on fertility preference in fodder collection models. The statistical significance disappeared when other individual-level factors such as exposure to media and perception of child raising costs were added. These results from firewood and fodder collection models suggest that gender has no independent effect on individual family size desires in this setting.

Next, we examined whether the effect of child labor use on individual fertility attitudes varies among men and among women by type of resource collection activities, *firewood,* and *fodder*. We expected that children's participation would equally contribute to larger family size desires of individuals in both of these activities and the results would be consistent with the results of pooled analysis discussed earlier. For the purpose, we performed further analysis by disaggregating gender and compared the net effects of child labor use within the same-sex groups, among men and among women. Models 1 through 3 in Tables 11.5 and 11.6 compare results among individuals within the same sex (among men or among women) that belonged to households that used child labor for collecting forest resources versus those who belonged to the households that did not use child labor. The effect of child labor use in shaping fertility attitudes without adjusting for other characteristics are presented in model 1, whereas the net effects of child labor use controlling for the effects of individual and household-level characteristics are presented in model 2 and model 3.

It is interesting to note that when we run gender disaggregated models, we found that the effect of child labor use differentially contributed to shaping fertility attitudes of women and men depending upon the type of activity children were involved in. Specifically, among men from households collecting firewood, those that used children preferred significantly larger family sizes ($\beta = 0.588$; $p<0.05$, model 6 in Table 11.5) as compared to those who were from households that did not use them. However, among women, though the direction of the effect was expected, the coefficient of child labor use was not statistically significant ($\beta = 0.175$; $p>0.10$, model 3 in Table 11.5). In contrast, among women of fodder-collecting households, those who used children preferred although weakly but significantly larger family sizes ($\beta = 0.273$; $p<0.10$, model 3 in Table 11.6) than those from households that did not use children, while the effect of child labor use was statistically non-significant among men (Table 11.6).

The gender aggregated analysis (pooled analysis) revealed that the use of child labor was important in shaping larger family size desires of individuals but the effect of gender *per se* was not important. The gender disaggregated analysis illustrated that the extent of the effect of child labor use in shaping fertility attitudes varies among men and among women and is contingent upon specific resource collection activities. For instance, while children's contribution to firewood collection was important in shaping larger

Table 11.5 Unstandardized OLS regression estimates predicting the effect of child labor use in firewood collection on the fertility preferences of currently married men and women of 15–45 years of ages with at least one living child

	Within sex group comparison					
	Women (n = 1146)			**Men (n = 862)**		
Variables	**Model 1**	**Model 2**	**Model 3**	**Model 4**	**Model 5**	**Model 6**
Independent variable						
Child labor use (used = 1)	0.778***	0.248	0.175	1.210***	0.662**	0.588*
Individual characteristics						
Age (years)	-	0.009	0.016	-	0.007	0.018
Number of living children	-	0.377***	0.362***	-	0.374***	0.344***
Child ever died (yes = 1)	-	0.694***	0.656***	-	0.072	0.032
Education (years of schooling)	-	-0.078***	-0.068**	-	-0.110***	-0.093***
Currently on salaried job (yes = 1)	-	-0.560	-0.638	-	0.152	0.089
Child raising cost (expensive = 1)	-	-0.100	-0.105	-	-0.233	-0.256
Exposed to FP on radio (yes = 1)	-	-0.277^{+}	-0.241	-	-0.204	-0.143
Household characteristics						
Cultivated land (kattha)	-	0.003	0.002	-	-0.002	-0.004
Toilet facilities (yes = 1)	-	-0.449**	-0.331*	-	-0.370*	0.229
Migration of for work (yes = 1)	-	-0.021	0.010	-	-0.051	-0.109
Distance to school (minutes)	-	-0.002	-0.003	-	0.000	0.000
Collection of firewood (non-foot = 1)	-	0.294	0.308	-	-0.199	-0.185
Alternative (adult) labor use						
Female labor (used = 1)	-	-0.209	-0.208	-	0.290	0.318
Male labor (used = 1)	-	0.556***	0.513**	-	0.162	0.105
Ethnicity (reference = others)						
Terai Tibetoburmese	-	-	0.519**	-	-	0.719***
Intercept	6.829***	5.700***	5.373***	6.405***	5.966***	5.383***
Model F	11.926***	16.228***	15.744***	23.683***	11.111***	11.286***
Regression degrees of freedom	1	15	16	1	15	16
Residual degrees of freedom	1144	1130	1129	860	846	845
Adjusted R-square	0.9%	16.6%	17.1%	2.6%	15.0%	16.0%

Note: FP: family planning programming 30 kattha = 1 hectare

*** = p<.001; ** = p<.01; * = p<.05; + = p<.10

Table 11.6 Unstandardized OLS regression estimates predicting the effect of child labor use in fodder collection on the fertility preferences of currently married men and women of 15–45 years of ages with at least one living child

	Within sex group comparison					
	Women (n = 1170)			Men (n = 865)		
Variables	**Model 1**	**Model 2**	**Model 3**	**Model 4**	**Model 5**	**Model 6**
Independent variable						
Child labor use (used = 1)	0.880***	0.282*	0.273+	0.888***	0.322+	0.298
Individual characteristics						
Age (years)	-	0.013	0.018	-	-0.002	0.008
Number of living children	-	0.318***	0.312***	-	0.401***	0.378***
Child ever died (yes = 1)	-	0.636***	0.613***	-	0.136	0.120
Education (years of schooling)	-	-0.066**	-0.056*	-	-0.081***	-0.065**
Currently on salaried job (yes = 1)	-	-0.651	-0.743	-	0.053	0.010
Child raising cost (expensive = 1)	-	-0.214	-0.214	-	-0.366+	-0.378+
Exposed to FP on radio (yes = 1)	-	-0.338*	-0.310+	-	-0.427+	-0.345
Household characteristics						
Cultivated land (kattha)	-	0.006*	0.005+	-	-0.003	-0.005
Toilet facilities (yes = 1)	-	-0.261+	-0.171	-	-0.625**	-0.487*
Migration for work (yes = 1)	-	-0.061	-0.039	-	0.011	0.053
Distance to school (minutes)	-	0.001	0.001	-	-0.004	-0.005
Collection of firewood (non-foot = 1)	-	-0.334	-0.316	-	0.070	0.111
Alternative labor use						
Female labor (used = 1)	-	0.276	0.341	-	0.296	0.390
Male labor (used = 1)	-	0.236+	0.214	-	0.030	-0.014
Ethnicity (reference = others)						
Terai Tibetoburmese	-	-	0.408*	-	-	0.754***
Intercept	6.485***	5.354***	5.030***	6.227***	6.631***	5.975***
Model F	37.576***	15.155***	14.533***	22.878***	9.288***	9.439***
Regression degrees of freedom	1	15	16	1	15	16
Residual degrees of freedom	1168	1154	1153	863	849	848
Adjusted R-square	3.0%	15.4%	15.6%	2.5%	12.6%	13.5%

Note: FP: family planning programming 30 kattha = 1 hectare
*** = p<.001; ** = p<.01; * = p<.05; + = p<.10

family size desires among men of firewood collecting households, their contribution to fodder collection was among women of fodder collecting households. Although women are also responsible for collecting firewood, it could be the fact that men share major responsibility in terms of time and effort in collecting firewood in the Valley. It may be because firewood is collected from the wildlife-infested dense forests of the Chitwan Valley from where women alone may not be able to collect and transport big logs. On the other side, fodder collection is a relatively easier task compared to firewood collection and therefore, women and children are particularly more involved in this activity. In this case, children may substitute women's work. It could be the reason why women from households that used child labor preferred significantly larger family sizes as compared to those who did not use them. The information on actual time involvement by women and men in firewood and fodder collection activities could have been helpful in further analyzing this issue, which is beyond the scope of this study.

The findings suggest that (a) although gender *per se* has no independent effect on individual fertility attitudes, children's contribution to environmental resources collection are important in shaping fertility attitudes among men and women, (b) however, the effect of child labor participation in shaping individual fertility attitudes among men and women is contingent upon the type of resources collection activity. In conclusion, gender as well as activity disaggregated analysis is required to understand the fertility attitudes and behaviors of individuals in environmental resources dependent communities in developing countries.

11.9 Effects of Other Factors

Individuals with larger number of children expressed larger family size desires. Child mortality experienced by individuals in their own lives was also a significant contributor to larger family size preferences (Tables 11.3 through 11.6). This result is consistent with the finding of Acharya (1998), who observed a positive relationship between child mortality and recent fertility in Nepal. The effect of education on fertility preference was statistically significant and negative as reported by Pearce (2000) and Gajurel (2001) in this research setting. Consistent with the findings of Westoff and Bankole (1991) in their study of Pakistan, India, and Bangladesh, the exposure to family planning programming on radio impacted significantly and negatively on individual fertility attitudes. As expected, the perception that "raising a child these days is costly" negatively and significantly contributed to smaller family size desires. Men who perceived "child raising is costly" preferred smaller family sizes than those who perceived "child raising is not costly." But the results are statistically significant among fodder collectors (Table 11.4), the results are in expected direction among firewood collectors (Table 11.3).

Among household characteristics, only the effect of the availability of toilet facilities in the home, a measure of socioeconomic indicator, had a statistically significant (except in men's model, Table 11.5) and negative impact on family size desires (Tables 11.3 through 11.6). This suggests that increased socioeconomic condition of a household negatively shapes fertility attitudes. However, although the effect of land size on fertility attitudes was in the expected direction as suggested by the land–labor demand framework (Stokes 1995; Stokes and Schutjer 1984 and Stokes et al. 1986), the coefficient is not statistically important. Similarly, migration of individual, presence of school nearby a household, and use of mechanical means of transporting environmental resources did not contribute significantly to shape individual fertility attitudes. By ethnicity, as reported in studies in this research setting, the local indigenous individuals belonging to the Terai Tibet Burmese group desired significantly larger family sizes as compared to the individuals of other ethnic groups.

11.10 Conclusions

- The incidence of hazardous work in countries affected by armed conflict is 50% higher than the global average.
- Nearly 1 in 10 children—some 152 million—are in child labor, almost half of whom work in hazardous conditions.
- Thirty million children live outside their country of birth, increasing their risk of being trafficked for sexual exploitation and other work.
- By 2025, an estimated 121 million children will be in child labor, with 52 million enduring hazardous work (UNICEF 2021—https://www.unicef.org/protection/child-labour).

As the scenario presented in the introduction section, the child labor is still one of the difficult challenges of all time, primarily due to the poverty, conflict, wars, and social scenarios. To minimize these challenges, it is necessary to be aware that children hold the future of the world and they have rights to enjoy their childhood and prepare for the future (Bhandari 2019; 2020).

In the specific case of Nepal, we investigated the effects of child labor use in two important environmental resources collection activities, *firewood* and *fodder*, and gender in shaping individual fertility attitudes using the individual and household-level data from the Western Chitwan Valley of Nepal collected in 1996. We further examined the effect of child labor use on fertility attitudes by disaggregating gender (men and women) and the effect of gender on fertility attitudes by disaggregating child labor use (those who used children and those who did not use them) to understand whether the effects are consistent among men and women or among child labor users and non-users, respectively. Overall, child labor contribution to environmental resources collection was found to be an important predictor of individual fertility attitudes. But the effect of gender *per se* was not important in shaping fertility attitudes. However, the gender disaggregated analyses suggested that the effect of child labor use in shaping men's and women's fertility attitudes appear to be affected by the type of activity adults are responsible for. For instance, while child labor use in firewood collection is important in shaping differential fertility attitudes among men, children's contribution to fodder collection is important in shaping differential fertility attitudes among women.

Based on the findings, we conclude that the use of child labor by households for environmental resources collection is an important factor contributing to shaping individual fertility attitudes. The three important messages that can be drawn from this study are: (i) individual fertility attitudes differ depending upon whether a household uses child labor or not in collecting environmental resources as against the common belief that fertility differentials exist between those who collect environmental resources and those who either do not use or do not collect (i.e., purchase from the market) them; (ii) gender *per se* is not an important predictor of individual fertility attitudes; (iii) however, the effect of child labor use in shaping fertility attitudes of men and women is contingent upon the type of activity children participate in. This study adds to the demography literature by providing evidence that variation in fertility attitudes exists among individuals within natural resources extracting communities depending upon the use or nonuse of child labor. This study also provides an understanding that family size desires vary among men and women depending upon the type of activities the households use child labor. These findings suggest a need for gender and an activity disaggregated analysis in fertility studies.

From a policy perspective, it can be suggested that in an agrarian society where child labor plays an important role in household economy, one of the ways to address high fertility problems may be to reduce the contribution of children to household activities and hence the demand for them. This needs an understanding of the questions such as why some households use child labor in certain activities and others do not? why children's participation in certain activities shapes larger family size desires among men, while participation in other activities shapes larger family size desires among women?, which are beyond the scope of this study. Further research investigating the intra-household labor dynamics, individual time allocation in various household activities and their relations to human fertility would provide added insight into these issues.

References

1. Acharya, Laxmi Bilas. (1998), "Determinants of Fertility in the 1970s and 1990s in Nepal." *Contribution to Nepalese Studies* 25 (Special Issue): 95–108.
2. Axinn, William G. and Jennifer S. Barber. (2001), "Mass Education and Fertility Transition." *American Sociological Review* 66:481–505.
3. Barber, J. S, G. Shivakoti, W. G. Axinn, and K. Gajurel. (1997). "Sampling Strategies for Rural Settings: A Detailed Example from the Chitwan Valley Family Study, Nepal." *Nepal Population Journal 6: 193–203.*
4. Basu, K. and H. Zarghamee (2005), "Is product boycott a good idea for controlling child labor?"
5. Basu, Kaushik (1999), Child labor: cause, consequence, and cure, with remarks on international labor standards. Journal of Economic literature 37.3 (1999): 1083–1119.
6. Becker, G. (1962), "Investment in human capital: A theoretical analysis", The Journal of
7. Becker, G. (1965), "A theory of the allocation of time", Economic Journal 75 : 493–517.
8. Becker, G. and H. Lewis (1973) "On the interaction between the quantity and quality of
9. Becker, Gary S. (1981), *A Treatise on the Family.* Mass: Harvard University Press.
10. Becker, Gary S. (1965), "A Theory of the Allocation of Time." *The Economic Journal* 75(299): 493–517.
11. Bhandari, P., M.R. Kolachchhapati, N.R. Devkota, P. Bhatt, M.Kharel, K. Basnet, M. Sharma and D. Baral. (1996), "Farming-Forestry and Environment: Socioeconomic Study of Makalu Barun Conservation Area Along Genderlines." A Research Report submitted to the Makalu-Barun Conservation Project (MBCP), Kathmandu, Nepal.
12. Bhandari, Medani P, (2019). Live and let other live- the harmony with nature /living beings-in reference to sustainable development (SD)- is

contemporary world's economic and social phenomena is favorable for the sustainability of the planet in reference to India, Nepal, Bangladesh, and Pakistan? Adv Agr Environ Sci. (2019);2(1): 37–57. DOI: 10.30881/aaeoa.00020 http://ologyjournals.com/aaeoa/aaeoa_00020.pdf

13. Bhandari Medani P. (2019). "BashudaivaKutumbakkam"- The entire world is our home and all living beings are our relatives. Why we need to worry about climate change, with reference to pollution problems in the major cities of India, Nepal, Bangladesh and Pakistan. Adv Agr Environ Sci. (2019);2(1): 8–35. DOI: 10.30881/aaeoa.00019 (second part) http://ologyjournals.com/aaeoa/aaeoa_00019.pdf
14. Bhandari, Medani P. (2020), Bashudaibha Kutumbakkam, all living being are our relatives, with Reference to journalist role on Corona Crisis-1, https://saralpatrika.com/content/31756/2020-05-30, Issue 31756, Kathmandu, Nepal
15. Bhandari, Medani P. (2020), Bashudaibha Kutumbakkam, All living being are our relatives, with Reference to journalist role on Corona Crisis-2, Issue 273/743 https://www.enepalese.com/2020/05/273743.html, Virginia, USA
16. Bhandari, Medani P. (2020), The Phobia Corona (COVID 19) - What Can We Do, Scientific Journal of Bielsko-Biala School of Finance and Law, ASEJ 2020, 24 (1): 1-3, GICID: 01.3001.0014.0769, https://asej.eu/resources/html/article/details?id=202946
17. Biddlecom, Ann E., William G. Axinn, and Jennifer S. Barber. (2000), Environmental Effects on Reproductive Preferences: A Case Study in Nepal. Paper presented at the Annual Meeting of the Population Association of America, Los Angeles, CA, March 23-25, 2000.
18. Bongaarts, John. (1978), "A Framework for Analyzing the Proximate Determinants of Fertility." *Population and Development Review* 4(1):105–32.
19. Caldwell, John C. (1982), "The Wealth Flows Theory of Fertility Decline." Pp. 333–51 in *Theory of Fertility Decline*, edited by John C. Caldwell. New York: Academic Press.
20. Caldwell, John C. and Pat Caldwell. (1987), "The Cultural Context of High Fertility in Sub-Saharan Africa." *Population and Development Review* 13:409–37
21. Chitrakar, P.L. (1990), *Planning, Agriculture and Farmers: Strategies for Nepal.* Publisher: Mrs. Ganesh Devi Chitrakar, Kathmandu, Nepal.
22. Coombs, Lolagene C. (1974), "The Measurement of Family Size Preferences and Subsequent Fertility." *Demography* 11(4):587–611.

23. Cunningham, Hugh, and Pier Paolo Viazzo (1996), Child Labor in Historical Perspective 1800-1985: Case Studies from Europe, Japan and Colombia. No. hisper96/1. UNICEF Innocent Research Centre.
24. Davis, Kingsley and Judith Blake. (1956), "Social Structure and Fertility: An Analytic Framework." *Economic Development and Cultural Change* 4 (4):211–35.
25. Easterlin, Richard A. (1975), "An Economic Framework for Fertility Analysis." *Studies in Family Planning* 6:54–63.
26. Easterlin, Richard A. and Eileen M. Crimmins. 1985. *The Fertility Revolution.* Chicago, London: The University of Chicago Press.
27. Edmonds Eric V. (2008), "Child Labor", Chapter 57 in T. Paul Schultz, John Strauss, Handbook of Development Economics, Volume 4. North Holland
28. Edmonds, Eric V. (2007), Child Labor, Nber Working Paper Series, Working Paper 12926, http://www.nber.org/papers/w12926, National Bureau of Economic Research, 1050 Massachusetts Avenue Cambridge, MA 02138, ttps://www.nber.org/system/files/working_papers/w12926/w12926.pdf
29. Edmonds, Eric, V., and Nina Pavcnik. (2005), «Child Labor in the Global Economy.» Journal of Economic Perspectives, 19 (1): 199-220. DOI: 10.1257/0895330053147895 https://www.aeaweb.org/articles?id=10.1257/0895330053147895
30. Fallon, Peter and Zafiris Tzannatos, (1998), Child labor : issues and directions for the World Bank /
31. Farooq, Ghazi M. (1985), "The Definition of Fertility: Measurement Issues." Pp. 125–48 in *Fertility in Developing Countries: An Economic Perspective on Research and Policy Issues,* edited by Ghazi M. Farooq and George B. Simmons. New York: St. Martin's Press.
32. Filmer, Deon and Lant Pritchett. (1997), "Environmental Degradation and the Demand for Children: Searching for the Vicious Circle." *World Bank Policy Research Paper Number 1623.* World Bank. Washington, D.C.
33. Gajurel, Kishor. (2001), Organization of Agricultural Production and Human Fertility. Unpublished Ph. D. Dissertation, The Pennsylvania State University, Pennsylvania.
34. Greene, M. E. and Ann E. Biddlecom. (2000), Absent and Problematic Men: Demographic Accounts of Male Reproductive Roles. *Population and Development Review* 26(1):81–115.
35. Gunnarsson, V., Orazem, P., Sanchez, M. (2006), Child labor and school achievement in Latin America. World Bank Economic Review 20, 31–54.

36. Gurung, O. P. 1987. Inter-relationship Among Pasture, Animal Husbandry and Agriculture: Case Study of Tara. National Resource Management Paper Series No. 2, HMG/USAID/GTZ/IDRC/FORD/ Winrock Project, Kathmandu.
37. Hilowitz, Janet. (2004), Child Labor: A textbook for university students. International Labor Office
38. His Majesty's Government of Nepal, National Planning Commission, Kathmandu, Nepal.
39. Humphries, J., (2010), Childhood and child labor in the British industrial revolution, Cambridge
40. ILO-IPEC, (2013), Marking progress against child labor – Global estimates and trends 2000-2012 / International Labor Office, International Programme on the Elimination of Child Labor (IPEC) – Geneva: ILO
41. ILO-IPEC, Diallo, Y., et al. (2013), Global child labor trends 2008 to 2012. ILO International Programme on the Elimination of Child Labor (IPEC). (Geneva, ILO, 2013).
42. International Labor Organization (1999a), C182 Worst Forms of Child Labor Convention, 1999. (ILO, Geneva).
43. International Labor Organization (1999b), R190 Worst Forms of Child Labor Recommendation, 1999. (ILO, Geneva).
44. International Labor Organization (2000), "Estimates and projections of the economically active population" in: Sources and Methods: Labor Statistics 10 (ILO, Geneva). International
45. International Labor Organization (2002), Every Child Counts: New Global Estimates on Child Labor (ILO, Geneva).
46. International Labor Organization (2006a), The End of Child Labor: Within Reach. (ILO, Geneva).
47. International Labor Organization (2006b), IPEC action against child labor: Highlights 2006. (ILO, Geneva).
48. Karan, Pradyumna and H. Ishii. (1996), *Nepal: A Himalayan Kingdom in Transition.* Tokyo, New York, Paris: United Nations University Press.
49. Kumar, Shubh K. and David Hotchkiss. (1988), *Consequences of Deforestation for Women's Time Allocation, Agricultural Production, and Nutrition in Hill Areas of Nepal.* International Food Policy Research Institute. Research Report 69, Washington, D.C.
50. Labor Organization (2001), The Time Bound Program in Nepal, (ILO, Kathmandu).
51. Lavalette, M (1999), A Thing of the Past? Child Labor in Britain in the Nineteenth and Twentieth Centuries. Liverpool University Press.

52. Loughran, David and Lant Pritchett. (1997), "Environmental Scarcity, Resource Collection, and the Demand for Children in Nepal." World Bank.
53. Mason, K. O and Taj, Anju Malhotra. (1987), Differences Between Women's and Men's Reproductive Goals in Developing Countries." *Population and Development Review* 13(4):611–38.
54. Mason, K. O. (1987), "The Impact of Women's Social Position on Fertility in Developing Countries." *Sociological Forum* 2:718–45.
55. Ministry of Health, Nepal; New ERA; and ICF. 2017. Nepal Demographic and Health Survey 2016.Kathmandu, Nepal: Ministry of Health, Nepal.
56. Montgomery, Mark R., Michele Gragnolati, Kathleen A. Burke, and Edmundo Paredes. (2000), Measuring Living Standards with Proxy Variables. *Demography* 37: 155–74.
57. NPC, (2003), The Tenth Plan: Poverty Reduction Strategy Paper 2002-2007. Summary. National Planning Commission, Kathmandu.
58. Pearce, Lisa D. (2000), The Multidimensional Impact of Religion on Childbearing Preferences and Behavior in Nepal. Unpublished Ph.D. Dissertation. Pennsylvania State University, USA.
59. Pebley, Anne R., Noreen Goldman, and German Rodriguez. (1996), "Prenatal and Delivery Care and Childhood Immunization in Guatemala: Do Family and Community Matter?" *Demography 33*: 231–247.
60. PRB. (2003), World Population Data Sheet of the Population Reference Bureau. The Population Reference Bureau, Washington, D.C.
61. Preston, Samuel. (1978), "Introduction." Pp. 1–18 in *The Effects of Infant and Child Mortality on Fertility,* edited by Samuel Preston. New York: Academic Press.
62. Robinson, Warren C. (1997), "The Economic Theory of Fertility over Three Decades." *Population Studies* 51:63–74.
63. Rosati, F., Rossi, M. (2003), Children's working hours and school enrollment: Evidence from Pakistan and Nicaragua. World Bank Economic Review 17, 283–295.
64. Rosenzweig, M. (1977), "Farm-Family Schooling Decisions: Determinants of the Quantity and
65. Rosenzweig, M. and Evenson, R. (1977), "Fertility, schooling, and the economic contribution of
66. Rosenzweig, Mark R. and Evenson, Robert. (1975), "Fertility, Schooling and the Economic Contribution of Children in Rural India: An Econometric Analysis." Yale Growth Center Discussion Paper no 239, November 1975.

67. Rosenzweig, Mark R. (1977), "The Demand for Children in Farm Households." *The Journal of Political Economy* 85(1):123-46.
68. Schultz, T. W. (1960), "Capital formation by education", Journal of Political Economy 68: 571-
69. Schultz, T.W. (2004), School subsidies for the poor: Evaluating the Mexican Progress poverty program. Journal of Development Economics 74, 199–250.
70. Sharma, Keshav Raj, and Rabi Prasad Kayastha. (1998), "Demographic and Economic Characteristics of Nepalese Population." Pp. 365–407 in A Compendium on Environment Statistics 1998, Nepal. Kathmandu: Central Bureau of Statistics, His Majesty's Government of Nepal.
71. Shivakoti, Ganesh P., William G. Axinn, Prem Bhandari and Netra B. Chhetri. (1999), "The Impact of Community Context on Land Use in an Agricultural Society." Population and Environment 20(3): 191–213.
72. Simmons, George B. (1985), "Research on the Determinants of Fertility." Pp. 67-106 in *Fertility in Developing Countries: An Economic Perspective on Research and Policy Issues,* edited by Ghazi M. Farooq and George B. Simmons. New York: St. Martin's Press.
73. Stokes, Shannon C and W. A. Schutjer. (1984), "Access to Land and Fertility in Developing Countries." Pp. 195–215 in *Rural Development and Human Fertility*, edited by Wayne A. Schutjer and Shannon C. Stokes. New York: Macmillan.
74. Stokes, Shannon C, Wayne A. Schutjer and Rodolfo A. Bulatao. (1986), "Is the Relationship Between Landholding and Fertility Spurious? A Response to Cain." *Population Studies* 40:305–11.
75. Stokes, Shannon C. (1995), "Explaining the Demographic Transition: Institutional Factors in Fertility Decline." *Rural Sociology* 60(1):1–22
76. Thapa, Shyam and Vinod Mishra, (2003), "Mass Media Exposure Among Urban Youth in Nepal". *Asia-Pacific Population Journal* 18(1):5–28.
77. IBRD (2004), Child Labor, The International Bank for Reconstruction and Development , Washington DC-
78. Thomas, Neil. (1991), "Land, Fertility, and the Population Establishment." *Population Studies* 45(3):379–97.
79. Thomson, E. and Y. Brandreth. (1995), "Measuring Fertility Demand." *Demography* 32:80–90.
80. Tuladhar, J. M. et al. 1982. "Differential Fertility in Rural Nepal." *Population Studies* 36(1):85
81. U.N. (1997), Levels and Trends of Fertility and their Determinants for Small Geographic Areas in the ESCAP Region. Economic and Social

Commission for Asia and the Pacific, Asian Population Studies Series No. 146. New York: United Nations.

82. Waite, Linda J. and Ross M. Stolzenberg. (1976), "Intended Childbearing and Labor Force Participation of Young Women: Insights from Non-recursive Models." *American Sociological Review* 41:235–52.
83. Westoff, C. F. and A. Bankole. (1999), "Mass Media and Reproductive Behavior in Pakistan, India, and Bangladesh." *Demographic and Health Surveys Analytic Report* Calverton, MD, USA: Macro International, No. 10.
84. WHO (1986), Children at Work: Special Health Risks, WHO, Geneva https://www.ilo.org/wcmsp5/groups/public/@dgreports/@dcomm/documents/publication/wcms_575499.pdf
85. WHO (2017), Global estimates of child labor: Results and trends, 2012-2016, International Labor Office (ILO), Geneva, 2017 ISBN: 978-92-2-130152-3 (print), ISBN: 978-92-2-130153-0 (web pdf)
86. World Bank (1995), World Development Report 1995. Workers in an Integrating World. New York: Oxford University Press.
87. World Bank (1996), "Thailand: Growth, Poverty and Income Distribution: An Economic Report". East Asia and Pacific Region, Report No. 15689-TH. Washington, D.C.
88. World Bank (1997), "Social Development and Results on the Ground." Task Force Report. Washington, D.C.
89. Yadav, Y. (1990), Farming-Forestry-Livestock Linkages in Mountain Regions. MFS No. 14. International Center for Integrated Mountain Development, Kathmandu, Nepal.

12

Conclusion—The Unpredictable Direction of Economic Inequality

Medani P. Bhandari and Shvindina Hanna

Economic inequality has been in the unpredictable direction throughout the economic development history. Economic status of the society depicts the real picture of public well-being. Economically unequal society remains at unrest and causes conflicts, violence, crime, and creates social deviance. It is a known truth that inequality has been a chronic divisive factor of society. It is a known truth/fact that several inequalities (economic, social, cultural, religious, geographical, etc.) have been omnipresent in human society. Inequalities can be found within each family, within a community, and from communities to the nation, and to other nations globally. Inequality is the major cause of political, economic, and social instability; it creates crisis and conflicts within the society and between nations. The major cause of inequality is the unequal, uneven, biased, power-centric distributions of economic, social, political, cultural, and spiritual necessities of humans. The human civilizations are grounded in a way that, a small elite group creates webs and networks within which they lead the society; rule the society in a way that makes the public rarely capable enough to get opportunities even to participate in fair competitions. Because in marginalized societies, individuals would not have similar access to wealth, education, health, and other benefits as the ruling elites have. Therefore, the elite groups remain elite for generations, and the dominated groups remain as victims. These scenarios can be seen as political inequality, which is the major cause of inequality in the society. The dominant group captures the political opportunities and power of governance so that marginalized group always remains at the lower strata of the political, social, and economic system. To minimize economic inequality; it is necessary to break the webs created by the traditionally inherited elites as well as by the newly emerged haves group; and empower, encourage, insist, and uplift the have-nots group, in a way that the dominated, marginalized group will also get equal opportunities to survive, grow, and enjoy their lives.

This book examines the major parameters of intertwined socio-economic issues of inequality and focuses on the key economic issues of inequality, namely, income and wealth distribution, equity & equality of outcome, and equality of opportunities. Basically, economic inequality is measured by wealth, income disproportions in distribution, and consumption patterns in the specific area. Mostly, for the statistical presentation, inequality is measured using Gini Coefficient, inequality adjusted human development index, 20:20 ratio, Palma ratio, Hoover index, Galt score, Coefficient of variation, Theil index, wage share, etc. However, it is not necessary that all income can be measured in the monetary unit and can be presented and proved through statistical tools. In this book, authors tried to unveil the problems and consequences of socio-economic inequality through case studies of different countries. Based on various case studies, this book encourages to rethink economic development through the lens of growing inequalities and disparities. These disparities became obvious with sharing information about global changes and distribution of environmental, economic, and social risks among the recipients. The economic growth per se is disproportional, and the efforts of scholars, practitioners, and policymakers should be directed to empower the marginalized society in a way that no one should left behind (UN Slogan). The main challenges and focus of every chapter are different, and yet in combination, they give an integrated understanding of the phenomenon of economic aspects of inequality.

Book also contains the forwards from prominent scholars of social sciences-academia, former staff of United Nations Agencies, and high ranking government officials who argue that "Lifting people out of poverty and reducing extreme income and wealth inequalities is important not only to provide for the dignity, well-being, and welfare of everyone in society, but also to ensure that the development remains sustainable. Inequality breeds conflicts, inefficiencies, and lack of security for all. It also is a moral hazard for the entire society" (forward). "In this complex, climate emergency, the most alarming challenge of 21st Century, if combined with global pandemic one ought to invent new economic thought of the day, which can work across all communities and societies and help improve universal prosperity, well-being, and improve quality of life. This means being sensitive to social values, ethics, economy, environment, and reduce excessive consumption that is depleting natural resources beyond replenishing, and not just work for glory, and serve privileged and wealthy. It is now time to put on-hold all excesses and insatiable thrust for infinite greed and dominance" (Forward).

This book tries to incorporate the alarming challenges of economic inequality with examples of specific area to community—national and international levels. The case studies presented in each chapter are unique in a sense that, they show the chain effects of how—disturbances in a particular scenario makes the global effect. Chapters also propose the new theories and new concepts to tackle the scenarios especially in the context of COVID impact in the development countries and in global context in general. Authors have introduced the new concept of CORONOMICS, drivers of sustainability, threats of development, the unpredictable direction of inequality among several other new and noble thoughts and arguments.

The current time is under the COVID crisis. This pandemic created complete uncertainty on human civilization and broke the chain of the social, political, economic, religious, and environmental system, which we have been practicing, from generation to generation. Now, we all are horrified, terrified, having phobia of uncertainty, wounded internally and externally, and even not able to prepare to make sequential efforts to fight with this pandemic. The late response or preparedness occurred because we could not foresee the connection of world and were not able to think that "the pain neighbor is facing could be my pain too." If that reality were realized on time, it would not have spread as it is now. Now there is no point of heating the wounded part. However, it is just in the beginning stage in many parts of the world; therefore, it is essential to take the protective measures as much as possible (the media has been publishing it every single hour—what to do—so it is not necessary to illustrate here—we all know—what to do). Most of us might be at home, still scared, frightened; however, what about the some of us who are in travel, who are trapped in different countries, or migrants, or the people in refugee camps due to inhumane greed of power. What if the pandemic penetrates in that already broken, dismantled, and sunk in sorrow frame of society, where people are dying every day without food and water? What about some of us who are still living in slums or the no nation situation? How will these come under the watch system? What will be the consequences of this phobic uncertainty? (Bhandari 2020).

It is a factual truth that COVID-19 has deepened and broadened economic, social, and political inequality in each corner of the planet, and the planet is wounded in a way that the healing will take a least a decade or so. When there is crisis, it is normal to lose patience, perseverance, and the mental strength, and there is always a chance of hopelessness. As such we live in hope and hope creates the motivation to move ahead. Motivation creates inner power in human heart as well as in the human brain. The current crisis has a direct impact to all of us and we are in the crisis of humanity, and even crisis of hope as well as crisis of motivation. Knowledge creates the motivation (Bhandari 2020).

Knowledge is the key to open the reality of the problem and knowledge helps to realize or to understand the facts, and wisdom is the proper utilization of acquired knowledge. As knowledge practitioners, we are aware that, we are by nature social beings and enjoy ourselves within the society. Society can only be harmonious when all members of the society maintain the societal norms and values. Society can only remain healthy—when we respect each other, encourage each other, and love each other and more importantly if we have balanced resources to maintain the life support system of the society. This notion applies to the large house as well "Bashudhaiva Kutumbakam." It is kind of an extended family, where all living beings share the same planet. Therefore, it is our responsibility, that each of the members can enjoy their part of resources. It is our responsibility to create the harmonious environment, where, live and let others live principle fully gets utilized (Bhandari 2018, 2019; 2020).

The contributors of this book have shown or chosen the notion of togetherness in addressing the challenges of inequality by contributing knowledge—from social science, environmental science, core science, or professional / vocational sciences point of views—which covers the spectrum of economic inequality. As we all know, knowledge itself does not solve the problem until it is implemented. The book covers inequality from a small geographical location (a case study) to the comparative views of different continents. This book presents a new knowledge, which tries to find the way to minimize the global challenges humanity is facing. The main message of the book is CORONA—a short-term pandemic; now, we have vaccine, and within a year or so, we will be protected. However, inequality is a long-term pandemic, a society divider; every one of us, all stake holders of this world, civil society, NGOs, the local community, districts, regions, governments, nations, and the global community, needs to join hands to create equity-based society, inclusive society, and united society so that the challenges of inequalities could be minimized.

Having this broad epistemology, the book presents the theoretical and practical examples of problems and consequences of economic inequality. Whereas, the first chapter provides the theory, general scenarios of inequality (based on World Inequality Report 2018), how inequality can be minimized (United Nations 2020; Walker 2006; Genevey et.al. 2013; Bhandari and Shvindina 2019), the economic and social burden is uneven (pictorial depictions); the tools of economic inequality measurement—Gini Coefficient, inequality adjusted human development index, 20:20 ratio, Palma ratio,

Hoover index, Galt score, Coefficient of variation, Theil index, wage share, etc. (World Inequality Report 2018; Berg and Ostry 2011; Freisteinand and Mahlert 2015; Guidetti 2014). This introduction chapter by Medani P. Bhandari, titled "the Problems and consequences of Economic Inequality" provides a general outline of the book and the issues addressed.

Chapter 2 by Man Bahadur Bk, titled Liberal Socialism: Path Forward to Reset the CORONOMICS, having the baseline information from Nepal, introduces the concept of CORONOMICS as a new way to see the contemporary development. This chapter examines how COVID has widened the inequality in the globe as well as in the developing world. Author argues that it has reversed the world economy creating the greatest recession around the globe since the 1930s. The pandemic that the world has been suffering is the fundamental issue, but there are multiple devastating effects on human lives and livelihoods tending toward a severe hunger epidemic. Through the experiences, it is justified that the neoliberal economic policy does not have adequate resilience. Capitalism has generated massive wealth for some, but it has devastated the planet and has failed to improve human well-being. So, the emergence of a new policy that balances economic mobility and human well-being is sought. The concept of "Protective Liberalism" can be a replacement to libertarianism as the socio-economic policy for "Liberal Socialism." It can also be coined as "Social Capitalism" that resets "Coronomics" in an egalitarian manner.

Likewise, Chapter 3, titled "On the Unified Index for Balanced Development and Inequalities Level Estimation" by Sergii Polumiienko (with the background of physics and mathematics), deals with an approach aimed at constructing of balanced development strategies for the country and its regions, including overcoming inequalities. There are defined regional (within the country) resources and development is considered as a set of system participants strategies aimed at creation of new resources using available information. The categories, types, and set of basic indicators of regional resources are introduced. The index creates good prospects for success in this work. An important area of research is to reduce data, including replacing them with available UN data and using them for expert analysis. At the same time, this information needs to be complemented by countries' regional data, not just using national data. This will allow to find real effective strategies for regional and further national development and for reducing inequalities.

Chapter 4 by Liubov Zharova and Natalia Eremeeva, titled Sustainable Development Goals as a Roadmap for Territorial Development, where authors argue that, the territorial development is a complex issue that could become the platform for overcoming after lockdown times and stimulate the economic growth of the country, because the seeking for a local solution could be simpler than reforming and stimulation of the growth on the governmental level. The government should create the framework based on the SDGs and territories would find their innovative way to prosperity (and Goals 7 can give them new understanding of resources and opportunities for development).

Similarly, Chapter 5 titled "Financialization as a Determinant of Income Inequality: The Case of Eastern European Countries" by Vasilyeva T., Lyeonov S., Pakhnenko O. and Rubanov P., analyzes the development of income inequality, driven by the growing role of the financial sector in the economy. The paper uses Eastern-European countries as a case study and provides a comparative analysis of the level of financialization in developed European countries and the Eastern European countries with lower levels of economic development, as well as a comparison of the impact of financialization on income inequality in the respective country groups.

Chapter 6 by Carlos Sanchez, titled "Poverty and Inequality in Access to Housing in Mexico City During Neoliberalism" provides a scalar explanation of the current conditions of poverty and inequality of the population of Mexico City caused by the lack of access to housing, understood as a chain of events unleashed by neoliberalism as the development path undertaken by Mexico from the eighties of the twentieth century: deindustrialization, depopulation, and poverty. While carrying out an exercise to measure and map poverty and inequality due to homelessness of the inhabitants of the 16 counties and functional areas of Mexico City as an exercise to find public policy responses to one of the most important challenges in the great metropolis.

Chapter 7, titled "Inequality in the Distribution of Income as a Threat to the Sustainable Development of United Territorial Communities in Ukraine, by *Ievgen Khlobystov and Lidiia Horoshkova*—analyzes effectiveness of intergovernmental fiscal relations system and income distribution within united territorial communities in the context of inequalities and imbalances as a threat to territories' and country's sustainable development. It has been proved that the current system of fiscal equalization of receipts and expenditure has some disadvantages related to basic and reverse subsidies. It has been proposed to determine the maximum allowable limits of reverse subsidy in relation to income growth rates per one UTC resident applying the logistical curve. Marginal value of the subsidy has been identified. Authors recommend to overcome existing disparities in UTCs budget expenditures distribution, it has been proposed to apply economic and mathematical model based on the first-order differential equation, which allows to consider simultaneously the dynamics of UTC per capita income and its population change. Application of the proposed models has practical significance as it provides opportunity to increase public policy efficiency to affect inequalities and imbalances as a principle for harmonious development of the country and its territories.

In Chapter 8, George Atisa begins with the "The North–South inequality hurts both high- and low-income countries in many respects, both directly and indirectly" and explains why it happens. In the chapter titled, A Moral and Pragmatic Analysis of Debt and North–South Economic Inequality, author tries to find answers of why do hard-working individuals in both the North and the South bring home different levels of income, yielding big differences in per capita income globally and what can be done about it? Current pro-poor growth policies that include handouts, loans, or relocating industries to the South cannot rid the world of inequality. The study argues that rich countries can and should prevent poor countries from sinking deeper into poverty if that does not call them to sacrifice anything of comparable moral importance. Data on GDP per capita and natural resources within countries is used to show and explain two forms of inequality; income levels and natural resources and that every region has something valuable to offer toward maintaining the global economy. The South is endowed with surplus natural resources that the North needs and the North has high incomes that can be used to support development needs in the South. Institutional weaknesses aside, the most feasible framework to reducing global inequality is a pragmatic and moral approach to valuing the natural resources from the South on the same scale as financial development assistance from the North.

Likewise, Chapter 9, titled "The Relationship Between Inequality and Socio-Economic Drivers as a Tool for Sustainable Territorial Development" by Yuliia Matvieieva, Yuliia Opanasyk, Olena Pavlenko, and Iuliia Myroshnychenko provides the classification of countries by the level of inequality between economic, social, and environmental areas of development based on *LPI*. According to the classification, there are countries with optimal equality, permissible inequality, and destructive inequality. For developing countries under conditions of sustainable development, *LPI* is in the range from 66.45 to 83.96, and for countries with destructive inequality, the *LPI* is in the range from 29.76 to 66.45. Thus, the equality between the economic, social, and environmental areas of the country's development and the permissible inequality between them within 70% lead to an increase in LPI, and, accordingly, to the country's prosperity in the context of the implementation of the sustainable development strategy. At the same time, the growth of inequality between these areas by more than 70% leads to a destructive transformation in the process of country's development.

Chapter 10 by Yuriy Petrushenko, Anna Vorontsova, and Oksana Ponomarenko, titled "Financial Inequality in Territorial Communities:

International Experience and Case-Study of Ukraine"—develops its own approach for estimation of the united territorial communities' financial capacity in Ukraine as an indicator of their financial inequality, based on the analysis of the current domestic and international methodologies. For this reason, the main stages and achievements of the decentralization process in the world and Ukraine were analyzed and the shortcomings of the existing methodologies for estimation of the united territorial communities' financial capacity were identified. The reform of decentralization is aimed at solving many problems existing at the local level due to the excessive centralization of power and its failure to meet the local needs of society and appropriately improve the quality of public services. Its first stage was focused on creating capable and effective administrative-territorial entities—united territorial communities; its consequences were analyzed in this chapter. The estimation of the financial capacity of the united territorial community is an important element of state regulation and management, which takes place both at the local and at the state level. Such results should be considered by the local authorities in defining the main priorities of their activities and in reducing possible financial inequality.

Chapter 11 titled "Child Labor Use in Forest Resources Collection and its Implications on Fertility Preferences in Nepal," by Prem Bhandari and Medani Bhandari, states that the advocacy and research about the child labor problem is not a new phenomenon in academia, as well as in developmental issues; however, the problem has not been resolved. This chapter addresses whether the children's involvement in forest resources collection contributes to having more children to maintain a household; however, the negative connotation to child growth has been always neglected. To solve the problem, it is necessary to have strong mechanism of policy implementation, and that is still way far. It is the role of every citizen that, the world remains in the shoulders of future generation and today's children are tomorrow's responsible citizens. Therefore, all concerned stakeholders should take responsibility to secure the future by nurturing today's child. Child labor has been utilized to perform various economic activities throughout the human civilization. Basically, poverty is the main cause of child labor; however, in some cases it can be just a traditional family practice. There are varieties of jobs in which children are involved where they should not be; however, this chapter primarily unveils the social reality of child labor use in forest resources collection and its implication on population growth. Using data from Nepal, we provide evidence that child labor in environmental resources collection contributes to a larger family size desire among individuals. Inequality of gender *per se* is not an independent predictor of individual fertility attitudes. The gender disaggregated analysis suggests that children's participation differentially shapes fertility attitudes among men and women depending upon the types of activities in which children participate. While children's contribution to firewood collection is important among men, their participation in fodder collection is important among women in shaping larger family size. The findings suggest a need for a gender and activity disaggregated focus to address high fertility problem in developing countries like Nepal. The chapter examines the effect of child labor use on fertility attitudes by disaggregating gender (men and women) and the effect of gender on fertility attitudes by disaggregating child labor use (those who used children and those who did not use them) to understand whether the effects are consistent among men and women or among child labor users and non-users, respectively. Overall, child labor contribution to environmental resources collection was found to be an important predictor of individual fertility attitudes. But the effect of gender *per se* was not important in shaping fertility attitudes.

The book chapters conclude that to minimize economic inequality; it is necessary to break the webs created by the traditionally inherited elites as well as from the newly emerged haves group; and empower, encourage, insist, and uplift the have-nots group, in a way that the dominated, marginalized group will also get equal opportunities to survive, grow, and enjoy their lives. And the main message of the book is that CORONA is a short-term pandemic; now, we have vaccine, and within a year or so, we will be protected. However, the inequality is a long-term pandemic, a society divider; every one of us, all stake holders of this world, civil society, NGOs, the local community, districts, regions, governments, nations, and global community, needs to join hands to create an equity-based society, inclusive society, united society so that the challenges of inequalities could be minimized.

Reference

1. Berg, Andrew, and Jonathan Ostry (2011). Inequality and unsustainable growth: two sides of the same coin? IMF Staff Discussion Note. SDN/11/08. Washington, D.C.: International Monetary Fund. 8 April.
2. Bhandari Medani P. and Shvindina Hanna (2019), The Problems and consequences of Sustainable Development Goals, in Bhandari, Medani P. and Shvindina Hanna (edits) Reducing Inequalities Towards Sustainable Development Goals: Multilevel Approach, River Publishers, Denmark / the Netherlands- ISBN: Print: 978-87-7022-126-9 E-book: 978-87-7022-125–2
3. Bhandari, Medani P. (2020), The Phobia Corona (COVID 19) - What Can We Do, Scientific Journal of Bielsko-Biala School of Finance and Law, ASEJ 2020, 24 (1): 1-3, GICID: 01.3001.0014.0769, https://asej.eu/resources/html/article/details?id=202946

4. Bhandari, Medani P, (2019), Live and let other live- the harmony with nature /living beings-in reference to sustainable development (SD)- is contemporary world's economic and social phenomena is favorable for the sustainability of the planet in reference to India, Nepal, Bangladesh, and Pakistan? Adv Agr Environ Sci. (2019);2(1): 37–57. DOI: 10.30881/aaeoa.00020 http://ologyjournals.com/aaeoa/aaeoa_00020.pdf
5. Bhandari Medani P., (2019), "BashudaivaKutumbakkam"- The entire world is our home and all living beings are our relatives. Why we need to worry about climate change, with reference to pollution problems in the major cities of India, Nepal, Bangladesh and Pakistan. Adv Agr Environ Sci. (2019);2(1): 8–35. DOI: 10.30881/aaeoa.00019 (second part) http://ologyjournals.com/aaeoa/aaeoa_00019.pdf
6. Freistein, K., and. Mahlert, B. (2015), The Role of Inequality in the Sustainable Development Goals, Conference Paper, University of Duisburg-Essen
   See discussions, stats, and author profiles for this publication at: https://www.researchgate.net/publication/301675130
7. Genevey R. (ed.), Pachauri R.K. (ed.), Tubiana L. (ed.), Jozan R., Voituriez T., Sundar S.. (2013). A planet for life: sustainable development in action, TERI, New Delhi http://regardssurlaterre.com/sites/default/files/edition/2016/01-PFL%202013%20Inequalities.pdf
8. Guidetti, Rehbein (2014), Theoretical Approaches to Inequality in Economics and Sociology, Transcience, Vol. 5, Issue 1 ISSN 2191-1150 https://www2.hu-berlin.de/transcience/Vol5_No1_2014_1_15.pdf
9. United Nations (2020), World Social Report 2020, Inequality in A Rapidly Changing World, Department of Economic and Social Affairs, New York https://www.un.org/development/desa/dspd/wp-content/uploads/sites/22/2020/02/World-Social-Report2020-FullReport.pdf https://www.un.org/development/desa/dspd/wp-content/uploads/sites/22/2020/01/WSR2020-Executive-Summary.pdf
10. United Nations (2020a), Inequality – Bridging the Divide- Shaping our future together- United Nations 75-2020 and Beyond, United Nations, New York https://www.un.org/en/un75/inequality-bridging-divide
11. Walker, M. (2006) Towards a capability-based theory of social justice for education policymaking. Journal of Education Policy, 21(2), 163-185.
12. *World Inequality Report (2018), The World Inequality Lab,* https://wir2018.wid.world/ https://wir2018.wid.world/files/download/wir2018-summary-english.pdf

Word Phrase Definitions and Abbreviation for Economic Inequality – Trends, Traps and Trade-offs

Abbreviation

COVID-19	Infectious disease caused by the strain of coronavirus SARS-CoV-2
CERF	UN Office for the Coordination of Humanitarian Affairs
EU	European Union
FAO	Food and Agriculture Organization
FIVIMS	Food Insecurity and Vulnerability Information and Mapping Systems
HIV	Human Immunodeficiency Virus
IBRD	International Bank for Reconstruction and Development (World Bank Group)
ISSC	International Social Science Council
IISD	International Institute for Sustainable Development Sustainable Development
ILO	International Labor Organization
IPCC	Intergovernmental Panel on Climate Change
IUCN	International Union for Conservation of Nature and Natural Resources
LGBTQ+	Lesbian, Gay, Bisexual, Transgender, and Queer and/or Questioning
MDG	Millennium Development Goal
MEA	Millennium Ecosystem Assessment
OCHA	UN Office for the Coordination of Humanitarian Affairs
ODI	Overseas Development Institute
PAP	Priority Actions Program
RAC	Regional Activity Centre

SA	South Asia
SD21	Sustainable Development in the 21st Century
SDGs	Sustainable Development Goals
UN	United Nations
UN DESA	United Nations Department of Economic and Social Affairs
UN ESCAP	Economic and Social Commission for Asia and the Pacific
UNAIDS	Joint United Nations Programme on HIV/AIDS
UNCCD	United Nations Convention to Combat Desertification
UNDP	United Nations Environmental Program
UNEP	United Nations Environment Programme
UNESCO	United Nations Educational, Scientific and Cultural Organization
UNHCR	Office of the United Nations High Commissioner for Refugees
UNHCR	UN High Commissioner for Refugees
UNICEF	United Nations Children's Fund
Vasudhaiva Kutumbakam	The entire world is my home and all living being are my relatives
WB	World Bank
WFP	World Food Program
WHO	World Health Organization
WWF	World Wildlife Fund

Word Phrase Definitions

20:20 Ratio It compares the ratio of the average income of the richest 20 per cent of the population to the average income of the poorest 20 per cent of the population. Used by the United Nations Development Programme Human Development Report (called "income quintile ratio") (United Nations 2015).

Atkinson's inequality measure (or Atkinson's index) This is the most popular welfare-based measure of inequality. It presents the percentage of total income that a given society would have to forego in order to have more equal shares of income between its citizens. This measure depends on the degree of society aversion to inequality (a theoretical parameter decided by the researcher), where a higher value entails greater social utility or willingness by individuals to accept smaller incomes in exchange for a more equal distribution. An important feature of the Atkinson index is that it can be decomposed into within and between-group inequality. Moreover, unlike other indices, it can provide welfare implications of alternative policies and allows the researcher to include some normative content to the analysis (Bellù, 2006) (United Nations 2015).

Climate Change (a) The Inter-governmental Panel on Climate Change (IPCC) defines climate change as "a change in the state of the climate that can be identified (e.g., by using statistical tests) by changes in the mean and/or the variability of its properties, and that persists for an extended period, typically decades or longer. Climate change may be due to natural internal processes or external forcing, or to persistent anthropogenic changes in the composition of the atmosphere or in land use" (IPCC).

Climate Change Includes both global warming and its effects, such as changes to precipitation, rising sea levels, and impacts that differ by region (IPCC).

Coefficient of Variation The coefficient of variation is the square root of the variance of the incomes divided by the mean income. It has the advantages of being mathematically tractable and its square is subgroup decomposable, but it is not bounded from above United Nations 2015).

CORONOMICS The Economic Analysis Based on change scenario due to Covid-19

COVID-19 Infectious disease caused by the strain of coronavirus SARS-CoV-2 discovered in December 2019. Coronaviruses are a large family of viruses which may cause illness in animals or humans. In humans, several coronaviruses are known to cause respiratory infections ranging from the common cold to more severe diseases such as Middle East Respiratory Syndrome (MERS) and Severe Acute Respiratory Syndrome (SARS). The most recently discovered coronavirus causes coronavirus disease COVID-19 (WHO, 2020).

Cultural Inequality discriminations based on gender, ethnicity and race, religion, disability, and other group identities. (ISSC, IDS and UNESCO 2016)

Decile dispersion ratio (or inter-decile ratio) It is the ratio of the average income of the richest x per cent of the population to the average income of the poorest x per cent. It expresses the income (or income share) of the rich as a multiple of that of the poor. However, it is vulnerable to extreme values and outliers. Common decile ratios include: D9/D1: ratio of the income of the 10 per cent richest to that of the 10 per cent poorest; D9/D5: ratio of the income of the 10 per cent richest to the income of those at the median of the earnings distribution; D5/ D1: ratio of the income of those at the median of the earnings distribution to the 10 per cent poorest. The Palma ratio and the 20/20 ratio are other examples of decile dispersion ratios (United Nations 2015).

Economic Inequality differences between levels of incomes, assets, wealth, and capital, living standards and employment. (ISSC, IDS and UNESCO 2016)

Empowerment Refers to increasing the personal, political, social, or economic strength of individuals and communities. Empowerment of women and girls concerns women and girls gaining power and control over their own lives. It involves awareness-raising, building self-confidence, expansion of choices, increased access to and control over resources and actions to transform the structures and institutions which reinforce and perpetuate gender discrimination and inequality (UNICEF 2017).

Environmental Inequality unevenness in access to natural resources and benefits from their exploitation; exposure to pollution and risks; and differences in the agency needed to adapt to such threats; (ISSC, IDS and UNESCO 2016).

Food Insecurity A situation that exists when people lack secure access to enough safe and nutritious food for normal growth and development and an active and healthy life. It may be caused by the unavailability of food, insufficient purchasing power, inappropriate distribution, or inadequate use of food at the household level. Food insecurity, poor conditions of health and sanitation, and inappropriate care and feeding practices are the major causes of poor nutritional status. Food insecurity may be chronic, seasonal, or transitory. (FIVIMS).

Food Security A situation that exists when all people, at all times, have physical, social and economic access to sufficient, safe and nutritious food that meets their dietary needs and food preferences for an active and healthy life. (FIVIMS)

Galt Score it measures the inequality of wage among workers. The Galt score is a simple ratio of a company's CEO pay to the pay of that company's Median worker. A company which pays its CEO many times more than its median employee will have a high Galt score (UNDP).

Gender A social and cultural construct, which distinguishes differences in the attributes of men and women, girls, and boys, and accordingly refers to the roles and responsibilities of men and women. Gender-based roles and other attributes, therefore, change over time and vary with different cultural contexts. The concept of gender includes the expectations held about the characteristics, aptitudes and likely behaviors of both women and men (femininity and masculinity). This concept is useful in analyzing how commonly shared practices legitimize discrepancies between sexes (UNICEF 2017).

Gender Analysis A critical examination of how differences in gender roles, activities, needs, opportunities, and rights/entitlements affect men, women, girls and boys in certain situations or contexts. Gender analysis examines the relationships between females and males and their access to and control of resources and the constraints they face relative to each other (UNICEF 2017).

Gini Coefficient Measure of the deviation of the distribution of income among individuals or households within a country from a perfectly equal distribution. A value of 0 represents absolute equality, a value of 100 absolute inequality (World Bank 2013).

Hoover Index Hoover index (also known as the Robin Hood index, Schutz index or Pietra ratio) It shows the proportion of all income which would have to be redistributed to achieve a state of perfect equality. In other words, the value of the index approximates the share of total income that has to be transferred from households above the mean to those below the mean to achieve equality in the distribution of incomes. Higher values indicate more inequality and that more redistribution is needed to achieve income equality. It can be graphically represented as the maximum vertical distance between the Lorenz curve and the 45-degree line that represents perfect equality of incomes (United Nations 2015).

Human Rights Agreed international standards that recognize and protect the inherent dignity and the equal and inalienable rights of every individual, without any distinction as to race, color, sex, language, religion, political or other opinion, national or social origins, property, birth, or other status (The Universal Declaration of Human Rights).

Inequality Unfair situation in society when some people have more opportunities, money, etc. than other people. (Cambridge Dictionary, and FAO, 2014).

Inequality-adjusted Human Development Index (IHDI) The IHDI combines a country's average achievements in health, education, and income with how those achievements are distributed among country's population by "discounting" each dimension's average value according to its level of inequality. Thus, the IHDI is distribution-sensitive average level of human development. Two countries with different distributions of achievements can have the same average HDI value. Under perfect equality the IHDI is equal to the HDI but falls below the HDI when inequality rises (UNDP).

Knowledge-based Inequality differences in access and contribution to different sources and types of knowledge, as well as the consequences of these disparities. (ISSC, IDS and UNESCO 2016)

Labor Binding labor law instruments include the 1973 Minimum Age Convention (No. 138), the 1999 Worst Forms of Child Labor Convention (No. 182), the 2011 Domestic Workers Convention (No. 189) of the International Labor Organization and the Protocol of 2014 to the Forced Labor Convention, 1930 (UNICEF 2017).

Palma Ratio It is the ratio of national income shares of the top 10 per cent of households to the bottom 40 per cent. It is based on economist José Gabriel Palma's empirical observation that difference in the income distribution of different countries (or over time) is largely the result of changes in the 'tails' of the distribution (the poorest and the richest) as there tends to be relative stability in the share of income that goes to the 'middle' (Cobham, 2015) (as in United Nations 2015).

Political Inequality the differentiated capacity for individuals and groups to influence political decision-making processes and to benefit from those decisions, and to enter into political action. (ISSC, IDS and UNESCO 2016)

Social Inequality differences between the social status of different population groups and imbalances in the functioning of education, health, justice, and social protection systems (ISSC, IDS and UNESCO 2016).

Spatial Inequality spatial and regional disparities between centers and peripheries, urban and rural areas, and regions with more or less diverse resources. (ISSC, IDS and UNESCO 2016)

Substantive Equality This focuses on the outcomes and impacts of laws and policies. Substantive equality goes far beyond creating formal legal equality for women (where all are equal under the law) and means that governments are responsible for the impact of laws. This requires governments to tailor legislation to respond to the realities of women's lives. Striving for substantive equality also places a responsibility on governments to implement laws, through gender-responsive governance and functioning justice systems that meet women's needs. Substantive equality is a concept expressed in the Convention on the Elimination of All Forms of Discrimination against Women (CEDAW). It recognizes that because of historic discrimination, women do not start on an equal footing to men (UNICEF 2017).

Theil Index Theil index and General Entropy (GE) measures the values of the GE class of measures vary between zero (perfect equality) and infinity (or one, if normalized). A key feature of these measures is that they are fully decomposable, i.e., inequality may be broken down by population groups or income sources or using other dimensions, which can prove useful to policy makers (United Nations 2015).

Wage Share The labor share is defined as the share of value added which is paid out to workers. It is therefore often also called the wage share. Generally, it is assumed that value added is produced with capital and labor as input factors so that $Y = F(K, L)$ where Y is value added or output1 , K the capital input, and L labor (Dorothee Schneider 2011).

Index

D

E

Estonia 91, 92, 100, 192

Q

R

T

U

V

W

Authors Biographies

Ambika P. Adhikari is a Principal Planner managing the long-range planning division at City of Tempe in Arizona, USA. He has over 30 years of professional and academic experience in urban and environmental planning and international development in several countries including Nepal, India, USA, Canada, Mexico, Kenya, and Fiji.

Ambika was a Village Planner and Project Manager at the City of Phoenix, and a Senior Planner at SRPMIC in Scottsdale, Arizona. He then joined Arizona State University as a Program and Portfolio Manager for Research. Prior to that, Ambika worked as a Director of international programs at DPRA Inc. in Toronto and Washington DC, where he managed international urban and environmental and economic development programs and projects in Canada, USA, Mexico, Nepal, and India.

In Nepal, he served as Nepal's Country Representative of the Switzerland based IUCN—International Union for Conservation of Nature, leading Nepal's national and some Asia regional programs in several environmental and conservation fields. He was also a member of the Government of Nepal's National Water and Energy Commission, the highest policy-making national body in this sector. Ambika was an Associate Professor of Architecture and Planning at the Institute of Engineering, Tribhuvan University, Nepal.

Ambika has authored one book, co-edited six books, published numerous reports and articles and refereed articles in journals. He regularly writes and lectures on sustainability and urban planning, international development, climate change policy, and international environmental policies and programs.

Ambika received his Doctor of Design (DDes.) degree in Urban Planning and Design from Graduate School of Design, Harvard University. He obtained his post-graduate fellowship from Massachusetts Institute of technology (MIT), M. Arch. from University of Hawaii, and B. Arch from M. S, University of Baroda.

He is a member of American Institute of Certified Planners (AICP) and holds the designations of Project Management Professional (PMP) through Project Management Institute (PMI), and LEED AP (ND) through the US

Green Building Council (USGBC). He is Fellow of American Society of Nepalese Engineers, where he also serves in the Advisory Council.

Ambika has been active in serving the Nepali community in the US and Canada by volunteering for various community organizations. He was the president of the NRNA National Coordination Council of US and Canada, NRNA Regional Coordinator for North America, Advisor to NRNA International Coordination Council (ICC), and Patron of NRNA ICC. He is the Chair of Board of Trustees for the Association of Nepalese in the Americas (ANA), Board Member of Asta-Ja USA, and Executive Member of Global Policy Forum Nepal (GPFN).

Anna Vorontsova graduated from the Ukrainian Academy of Banking of the National Bank of Ukraine (M.S in "Accounting and Auditing"), Sumy State University (Ph.D. in "Money, finance and credit"). Currently works as a senior lecturer and senior researcher of the Department Of International Economic Relations at Sumy State University (Ukraine). She is the author of more than 60 papers, was awarded the scholarships of the Cabinet of Ministers of Ukraine for young scientists (2017). Scientific interests include Educational policy, Lifelong learning system, National and Local development. She is a member of NGO "Council of Young Scientists" and NGO "Lifelong learning centre."

Douglass Lee Capogrossi has established and administered a variety of formal and non-formal education programs including distance learning colleges, trade apprenticeships, correctional education programs, work experience projects, on-the-job training ventures within industry and the human services, cross-border university affiliations, distance learning training programs for industry, and adult job training through center-based programs. He holds permanent teaching credentials in the United States in commerce and social studies and lifetime certification as a counseling teacher for the emotionally disturbed. Dr. Capogrossi has ten years' experience developing and delivering successful correctional education programs for adult inmates in Hawaii prison facilities with emphasis in parenting, cognitive development, transition to work and community, and adult basic education skills. Dr. Capogrossi is an experienced community service administrator, where his expertise rests primarily with creation of NGO corporate structures, program funding, excellence of Board operation, and implementation of a wide spectrum of government and nonprofit training programs and emergency services projects. Dr. Capogrossi has held top management positions in industry, serving as General Manager of Micrographic Systems, a medical camera-manufacturing firm in Silicon Valley California for a brief period in the early 1980s and he owned and operated America Builders, a successful licensed general contracting firm. Dr. Capogrossi has extensive community service experience on NGO boards and is highly experienced with founding activities for new nonprofit ventures, especially with antipoverty programs. He has served as a member of the Board of an international quality assurance agency in higher education. Dr. Capogrossi earned his bachelor's in business administration, his Master's in Curriculum and Instruction, and his Ph.D. in Adult and Continuing Education from Cornell University, USA, where he completed an extensive dissertation investigating the effectiveness of the American education system. Among his more recent published scholarly papers, The Assurance of Academic Excellence among Nontraditional Universities was published in the Journal of Higher Education in Europe. Dr. Capogrossi has dedicated his career in service to humanity, through his efforts to improve the human condition.

Dr. Durga D. Poudel, the Founder of Asta-Ja Framework and Professor in School of Geosciences at University of Louisiana at Lafayette, Lafayette, USA, was born and raised in Tanahu, Nepal. He spent his childhood in Tanahu and Lamjung districts, both in the Mid-Hills of Nepal. Dr. Poudel passed his S.L.C. exam from Nirmal Vocational High School, Damauli, Tanahu in 1977 and received an I.Sc. in Agriculture at Tribhuvan University at Rampur (IAAS), Chitwan, Nepal in 1980; a B.Sc. in Agriculture (Major: Agricultural Economics) at University of Agriculture, Faisalabad, Pakistan in 1987; a M. Sc. in Natural Resource Development and Management at Asian Institute of Technology, Bangkok, Thailand in 1991; and a Ph.D. in Soil Science, University of Georgia, Athens, GA, U.S.A. in 1998. Dr. Poudel is an expert in environmental science, climate change adaptation, and sustainable agricultural development; soil physical, chemical, and mineralogical characterization; soil classification; soil and water conservation, water quality, roadside vegetation management, and natural resources conservation and development. His recent research focuses on water quality monitoring and modeling, smallholder mixed-farming system, climate change adaptation, geohazards, wildflower, and environmental soil chemistry. Dr. Poudel has published over 100 peer-reviewed journal articles, conference proceedings papers, book chapters, research briefs, and scientific abstracts. He has published numerous popular articles. He has given over 100 scientific presentations to regional, national, and international conferences. Dr. Poudel has received more than $4.48 million in external funding as a PI and more than $1.76 million in external funding as a CO-PI.

He has supervised more than two dozen graduate students in their Thesis/Dissertation research.

Dr. Poudel's professional experience consists of Research Fellow at Asian Vegetable Research and Development Center, Taiwan (1991–1994); Graduate Research Assistant in Sustainable Agricultural and Natural Resource Management Collaborative Research Support Program, University of Georgia (1994–1998); and Visiting Research Scholar, University of California Davis (1998–2000), USA. Dr. Poudel joined the University of Louisiana at Lafayette, USA, as an Assistant Professor of Soil Science in August 2000, and currently is a tenured Professor and Coordinator of Environmental Science Program of School of Geosciences, Director of Ag. Auxiliary Units (Model Sustainable Agriculture Complex (600-acre Cade Farm), Crawfish Research Center, and Ira Nelson Horticulture Center), and Board of Regents Professor in Applied Life Sciences at the University of Louisiana at Lafayette, Louisiana, USA. Dr. Poudel's professional affiliations include membership in Soil Science Society of America, American Society of Agronomy, Crop Science Society of America, Geological Society of America, American Geophysical Union, and Soil and Water Conservation Society, USA. Dr. Poudel is the life member of NRNA, Nepalese Association in Southeast America (NASeA), Asta-Ja Abhiyan Nepal, Asta-Ja USA, and Association of Nepalese Agricultural Professionals of Americas (NAPA). Dr. Poudel is the founding member of Asta-Ja Abhiyan Nepal, and the founding President of Asta-Ja Research and Development Center (Asta-Ja RDC), Kathmandu, Nepal, and Asta-Ja USA, Honolulu, Hawaii, USA. Dr. Poudel was the Chair of NRNA ICC Agriculture Promotion Committee America Region, 2017–2019. Dr. Poudel served as a Board Member of the Bayou Vermillion Preservation Association (BVPA), Lafayette, Louisiana, USA, and an Advisor of the NASeA. Dr. Poudel is a Resource Person of Louisiana Organics, and a member of the Louisiana Technical Advisory Committee, USDA-NRCS, Louisiana, USA. In 2017, in recognition of the impact and quality of one of his research papers, Dr. Poudel and his co-authors were awarded for the "2017 Best Research Paper Award for Impact and Quality Honorable Mention" by the Journal of Soil and Water Conservation. Dr. Poudel believes on providing his services to others, promoting social inclusion, integrity, goodwill, and peace, and advancing on better understanding of the world.

George Atisa is an Associate Professor at the University of Texas Rio Grande Valley in Texas. He teaches in the Master of Public Administration in the department of Public Affairs and Security Studies. Dr. Atisa has a PhD in

Public Affairs and an MSc in Environmental Sciences both from Florida International University in USA, an MA in Economics from the University of Nairobi and a BA from Kenyatta University in Kenya. His research and publications span across governance, public management, political and administrative decentralization, environmental management, and sustainability. Prior to coming to academia, Dr. Atisa worked for WWF-International, Lutheran World Federation and International Rescue Committee. He is currently a Commissioner with the Commission on Environment, Economics, and Social Policy of the IUCN. He is skilled in program planning, design, operations, and evaluation.

Hanna Shvindina graduated from Sumy State University (M.S. in "Management of Organizations, Ph.D. in "Environmental Economics and Natural Resources Protection," Dr.Sci. in "Economics & Management at the Enterprises"). She was awarded several individual scholarships and was a Post-Doc researcher in France in the US (University of Montpellier, Purdue University) in "Business Studies and Management," Fulbright Alumna (2018–2019), won ERASMUS MUNDUS scholarships (2014, 2015). Currently, she works as a Head of the department and Associate Professor at the Department of Management at Sumy State University (Ukraine). She is a member of CENA community, Researchers' excellence network (RENET), Ukrainian Association for Management Development and Business Education. She is a member of editorial boards of several scientific journals (Ukraine, Poland, Switzerland), invited reviewer for the international conferences (Poland, Germany, Ukraine). She is the author of more than 90 papers. Scientific interests include Coopetition Paradox, Strategies and Innovations, Change Management, Organizational Development, Civic Education and Leadership. She is an activist and is a CEO at NGO "Lifelong learning centre," where provides courses on Leadership and Communications, organizes the program on empowerment for the local community, and offers advisory services.

Hemant Ojha works partly as Associate Professor for University of Canberra and partly as a Principal Advisor for Sydney based research and development consulting firm *Institute for Study and Development Worldwide (IFSD)*. Dr Ojha's work cut across academic theorizing, social entrepreneurship, and public policy engagement in relation to social development and environmental sustainability. A PhD on environment and development from University of East Anglia, Dr Ojha is a well-recognized public intellectual on Nepal's community forestry which is seen as a global institutional innovation for tackling the twin challenges of environment and development. He has championed science-policy interface works in South Asia in the field of forestry, agriculture and climate change. He founded and led various policy think tanks and education institutions in South Asia and Australia. He has taught at Australia's University of New South Wales and has conducted research in Asia and Africa on a range of environmental governance issues. His current research focuses on social entrepreneurship, public leadership, and science-policy interface in sustainable development. His Books have been published by Routledge and Cambridge and writes actively in both academic and public media.

Horoshkova Lidiia, chief researcher of the Department of Environmental Studies of National University of Kyiv-Mohyla Academy (NaUKMA), Kyiv, Ukraine. Candidate of Physical and Mathematical Sciences (01.04.07—solid state physics) and Doctor of Economics (08.00.03—Economics and Management of the National Economy), Associate Professor. Author of more than 260 scientific papers, including 2 textbooks (Production Management and Entrepreneurship, Trade, Exchange Activities (section)), 9 monographs (including 2 in the EU languages), 4 patents, 4 registration of the copyrights certificates; articles indexed in scientometric databases Scopus and WoS; 5 scientific and methodical publications. h-індекс 2 by Scopus; h-індекс 3 by WoS2019 (32 + 1 citations). A member of two specialized scientific councils (doctoral) at Azov State Technical University and Yuri Kondratyuk Poltava National Technical University, a member of the editorial boards of three professional journals in Ukraine, indexed in international scientometric databases. Since 2016—Scientific Secretary of Section 22 «Earth Sciences» of the Scientific Council of the Ministry of Education and Science of Ukraine. Since 2020—expert of National Research Foundation of Ukraine (economic geography, economic geology, economics, and management of the national economy). Research interests: economic and mathematical modelling of complex natural and socio-economic systems, economics and management of the national economy, forecasting, sustainable development, environmental management, decentralization of power, logistics, economic geology, economic geography, land management and cadastre, environmental protection.

Ievgen Khlobystov is a Doctor of Economic Sciences, Professor, Laureate of the State Prize of Ukraine in Science and Technology, academician of the Academy of economic science of Ukraine. Started his academic career at The Council for Studies of Ukrainian Productive Forces, National Academy of Science of Ukraine. He is an international expert in environmental policy and sustainable development. Lately Ievgen Khlobystov has been teaching at the Department of Ecology at the National University "Kyiv-Mohyla Academy" and at the University of Economics and Humanities (Bielsko-Biala, Poland).

Iuliia Myroshnychenko graduated from Sumy State University with a master's degree in "Management of Organizations" and a PhD in "Economics and management of enterprises." Currently, she works as an Associate Professor at the Oleg Balatskyi Department of Management, Sumy State University (Ukraine). Post-Doc researcher in University of Porto (2016–2017) and Vienna University of Economics and Business (2018–2019), the participant in the EMINENCE project for academic staff mobility at the University of Montpellier (October 2015). She is a Ukrainian Association member for the Development of Management and Business Education (UARMBO) from 2020. Iuliia Myroshnychenko is a member of editorial boards of several scientific journals (Ukraine, United Kingdom, USA) and has published numerous articles in scientific journals. Her research focuses on investment and innovation, the modelling of macro-and microeconomic processes, corruption, and institutional quality.

Dr. Jacek P. Binda is a Recotr and a researcher at Bielsko-Biała School of Finance and Law, Poland (pol. Wyższa Szkoła Finansów i Prawa w Bielsku-Białej, Polska). Prof. Binda obtained his PhD at the Warsaw University of Technology and his Post-Doc (habilitation) at University of Zylina, Republik of Slovakia. His principal fields of research include economy and finance, problems of the high-risk financial instruments, banking, local and public finances, e-economy, local government, project management and controlling. Since 2005 he has coordinated several medium-size and large research projects focusing on border studies, supported by the EU's Framework Programmes, National Research and Development Centre in Warsaw, Poland. He is an expert of the National Research and Development Centre in Warsaw, European Commission, Scientific Grant Agency of the Ministry of Education of Slovak Republic and the Academy of Science. Author or co-author of numerous monographs, over 60 national and international publications in the field of finance and information technology. He is a member of the Scientific Boards and Editorial Boards, incl.: Strategic Planning for Energy and the Environment (Journal in River Publishers), MDPI Journals, Switzerland; Sumy National Agrarian University: Sumy, UA; Scientific Journal of Bielsko-Biala School of Finance and Law (Editor-in-Chief), PL. Awarded the Gold Cross of Merit for achievements in research and teaching by President of Poland.

Jagadish Timsina is a systems agronomist specializing on crop, soil, nutrient and water management; conservation agriculture based sustainable intensification; agroforestry and community forestry systems; and crop and systems modelling. He has diverse experience in agronomic, economic, and environmental management of agro-ecologies ranging from levelled lands in Australia to hills and mountains in South and Southeast Asia. He has worked with University of Melbourne and Commonwealth Scientific and Industrial Research Organization (Australia), International Rice Research Institute (Bangladesh and Philippines), International Maize and Wheat Research Centre (Bangladesh and Mexico), and Tribhuvan University and Agricultural and Forestry University (Nepal) for about 40 years. Currently, he is an editor to Agricultural Systems journal, a Senior Fellow at the Global Evergreening Alliance (Melbourne), an organization dedicated to land and forest restoration and drawing down of greenhouse gases globally, and a Senior Adviser to the Institute for Study and Development Worldwide (Sydney).

Kedar Neupane holds M.A. (Economics) from Tribhuvan University of Nepal and a Rockefeller Fellow Scholar at the School of Economics of University of the Philippines, was trained as emergency preparedness and operations manager at the University of Wisconsin (USA). After retiring from work for over three decades of international service with the United Nations System Mr. Neupane currently resides in Geneva (Switzerland). He began international professional career starting from the UN Department of Technical Cooperation and Capital Development Fund, attached to the UNDP Sudan, prior to moving to UNHCR. Mr. Neupane has worked for the United Nations High Commissioner for Refugees in various capacities rotating between several countries in the North-Africa, the Middle East, Africa, Asia, and Europe. With varied work experience acquired from several countries, Mr. Neupane has unique perspectives on developmental issues and operational challenges, to bottom-up approach to the people-oriented program, design, planning, and implementation with target beneficiaries at the center of operational complexity of development syndromes, financing, and strategies for sustainability, organizational change management, and reform.

Dr. Keshav Bhattarai, PhD, Born and raised in a rural area of Balkot (Current Chhatradev Village Palika), Arghakhanchi district of Nepal, Dr. Keshav Bhattarai has contributed hugely to bridge Nepali and US education system.

Bhattarai family comprised of four siblings (three brothers and one sister). The family lived on subsistence farming. Bhattarai completed High School

(10th grade) from Mahendra Vidya Bodh High School, Balkot, Arghakhanchi district of Nepal. He earned Intermediate of Science (I. Sc.), Bachelor of Science (B. Sc.), and Bachelor of Arts (BA) from Tribhuvan University. He also earned Associate of Indian Forest College (AIFC) post-graduate Diploma from Indian Forest College, Dehradun, India and Master of Science (M.Sc.) in Natural Resource Management from the University of Edinburgh, Scotland, U.K.

He completed Ph. D. from Indiana University, Bloomington, Indiana, US in 2000 in Geography. He later joined Eastern Kentucky University as a faculty member, and in 2002, he moved to the University of Central Missouri, Warrensburg, Missouri.

His career on academia began as an Associate Instructor at the Department of Geography, Indiana University, Bloomington, Indiana in 1996. He has since taught a variety of courses at Eastern Kentucky University and University of Central Missouri. Dr Bhattarai is a member of Nepal Foresters' Association since 1983 and Association of American Geographers since 2002. A recipient of more than a dozen grants, Dr. Bhattarai has also taken the responsibility of journal article reviewer and book reviewer.

Dr. Bhattarai has been working as a Professor/ Geography at the School of Geoscience, Physics, and Safety Sciences at the University of Central Missouri since 2011. Dr. Bhattarai has also worked as a Short-term Consultant to the World Bank to assess Hydropower potential in Nepal. He has also worked as an Interim Chair, Department of Geography at the University of Central Missouri from 2009 to 2011.

Dr. Bhattarai bridges between the University of Central Missouri (UCM) and various universities of Nepal. Nepali universities and UCM have started 2+2 and 3+1 programs with UCM. The 2+2 refers to completing two years of undergraduate studies in Nepal and remaining 2 years in the US to complete undergraduate studies and vice-versa. The 3+1 program refers to completing three years education in Nepal and transferring to the US and vice-versa to complete an undergraduate degree.

Before embarking for his higher studies in the USA, Dr. Bhattarai worked in various capacities under the Ministry of Forest and Soil Conservation in Nepal from 1983 to 1995. Bhattarai left the Government job due to repeated political illogical impositions on him on the settlements of political cadres on forest areas without any norms and land use planning.

Dr. Bhattarai's urbanization research reveals that Nepal's urbanization is moving too quickly without needed infrastructure for the urban dwellers. The unplanned political ad-hoc decision to annex many rural areas into urban definition has increased vulnerabilities among urban dwellers since Nepal is located in between frequently moving Eurasian and Indian tectonic plates. Nepali urbanization has become *ruralopolis* (political decision to annex rural areas into municipalities). Such ad-hoc urbanization has increased the costs of lands and taxes on low-income people without giving any needed facilities to them. Dr. Bhattarai served as Fulbright Specialist at the Central Department of Geography (CDG) at Tribhuvan University, Kathmandu in the summer of 2018 to facilitate CDG to embark to active participation in urban planning through research.

Dr. Bhattarai is the recipient of Gorkha Dakhsina Bahu, award for civil servant decorated by the King Birendra for outstanding performance in forestry service and administration in 1993. He has also received Mahendra Vidyabhusan for his academic excellence in 2004. Moreover, he is also the recipient of All Round Forest Award by Indian Forest College, Dehradun, India in 1983, International Scholar Award by the University of Central Missouri in 2003 and Chairperson's award for Outstanding Academic Performance by the Department of Geography, Indiana University, Bloomington in 1999. He has also bagged half a dozen faculty research award for his academic excellence from 2007 to 2020.

He has published a number of academic papers, authored four books, co-authored one book and contributed many book chapters, journal articles, and presented his research at a number of professional meetings.

Khlobystov Ievgen, professor of the Department of of Environmental Studies, National University of «Kyiv-Mohyla Academy» (NaUKMA), Kyiv, Ukraine. Doctor of economic science (08.00.06—environmental economics and environmental protection, professor. Started his academic career at The Council for Studies of Ukrainian Productive Forces, National Academy of Science of Ukraine. He is an international expert in environmental policy and sustainable development. Author of more than 300 scientific and scientific and methodical publications, among them: 44 monographs (one individual, others co-authored), 4 textbooks. Now he is a member of the Specialized academic councils of Kyiv National Economic University named after Vadym Hetman and Sumy State University. From 2010 to 2016, he was an expert of the expert council for awarding scientific degrees and conferring academic titles in Economics of the Supreme Certifying Commission of Ukraine of the Ministry of Education and Science of Ukraine. Member of the editorial boards of several journals in the field of research. He was supervisor of 14 Ph.D.s and 2 Doctors of Science. Research interests: sustainable development, environmental management, decentralization of power, environmental protection.

Krishna Prasad Oli, currently member of National Planning Commission, Government of Nepal comes from Tehrathum District, Province No.1 of Nepal. Dr. Oli has the experiences of working with National and International organizations in a wide range of development and environmental issues at policy–program-action continuum in particular to environment planning and management, ecosystem management, implementation of multilateral environmental treaties, developing regional cooperation mechanism, capacity building, trans-boundary landscape management, and policy coordination and networking with stakeholders at different levels. He is a member of Nepal Bar Association and before joining the Government he was a visiting faculty at TU, teaching Environmental Law, he was Adjunct Professor at Faculty of Social Development and Western China Development Studies in Sichuan University in China and Distinguished Professor at Xian Minzu University, China. He is also a visiting professor at Kathmandu School of Law and adjunct faculty at Akamai University, Hawaii, USA. He has extensively visited different parts of Nepal, China, South and South East Asia, Europe, Canada, North America, Australia and Africa during his career.

Lidiia Horoshkova is a Professor of the Department of Business, Management of Organizations and Logistics of Zaporizhzhia National University. Professional degree: Ph.D. in Physical and Mathematical Sciences (1995), Doctor of Economic Sciences (2014). Academician of the Academy of Economic Sciences of Ukraine. Scientific secretary of Section 22nd "Earth Science" of the Scientific Council of the Ministry of Education and Science of Ukraine. Sphere of scientific interests: economic and mathematical modeling of complex nature-socio-economic systems, economy and management of national economy, sustainable development, environment economics, decentralization of power, logistics: h-index 2 by Scopus, 3 by WoS.

Liubov V. Zharova is a Doctor of Science in Economics, Prof. Wyższa Szkoła Ekonomiczno—Humanistyczna (Poland), Head of the department of Ukrainian American Concordia University, worked with the State Fund for Fundamental Research of Ukraine, the Ministry of Ecology and Natural Resources of Ukraine, the Ministry of Economic Development, Trade and Agriculture of Ukraine, the European Bank for Reconstruction and Development, and the UNDP. L. Zharova obtained a Ph.D. in Economics ("Productive Forces and Regional Economics") in 2005 and was awarded the academic title of Senior Researcher in "Economics of Nature Management and Environmental Protection" in 2007. In 2013, she defended her doctoral research thesis "Macroeconomic regulation of environmental policy:

theory, methodology, practice" and was awarded a Doctor of Science degree in Economics ("Economics of Nature Management and Environmental Protection"). Sphere of professional interests' sustainable development, green economics, macroeconomics trends, and governance.

Man Bahadur Bishwakarma (BK) has completed his postdoc research on food security as a Fulbright Visiting Scholar (2016/17) at Brandeis University, USA. He obtained Ph.D. on Social Inclusion in Microfinance in 2010 from Tribhuvan University, Nepal. He did a postgraduate diploma in Social Studies from the ISS, Netherlands (2004), MBA (1991) in Financial Management and MA (Economics) in Rural and Cooperative Development (2000) from Tribhuvan University, Diploma (1997) in Agro-Cooperative Management from Japan. He has long working experience with various international development agencies such as GTZ, Save the Children, UNICEF, CWS UK, UNDP. He has profound experience in conflict, peace, and the constitution building process as he led the Constitutional Dialogue Centre/UNDP during the Constitutional Assembly-I. In 2010, he rejoined the government bureaucracy as a joint secretary and served as the Director General for Labour Department, Chief District Officer in several districts, Secretary for provincial Government of Lumbeni and Karnali; and now the Principal Secretary for Bagamati province. He has been serving as a visiting faculty member for the M.Phil Program (Education) and MBA-BF Program of Tribhuvan University and written a dozen of textbooks for college students and several books on development issues, including Social Inclusion in Microfinance, Eradicating Hunger: Rebuilding Food Regime.

Dr. Mary Jo Bulbrook, her career expansion covers over 51 years primarily serving higher education in leadership positions in academia starting in community college education in nursing, then higher education at Texas Woman's University to University of Utah. Then she was guided to move out of country to Memorial University of St. John's Newfoundland, Canada, and from there to Edith Cohen University in Perth, Western Australia. On the side her role was a practitioner, educator, and researcher in a range of energy therapies starting with Therapeutic Touch, Healing Touch, Touch for Health, Transform Your Life through Energy Medicine (TYLEM) her own creation blending her profession as a psychiatric mental health clinical specialist with energy therapies. This expansive background of energy therapies was not only taught throughout USA and Canada but also taken to Australia, New Zealand, South Africa, and Peru where she worked with traditional healers while education professors, nurses, and other health care, and lay people.

Her journey served to capture the worldwide perspective on education, health and healing to address local, national and international ways to improve the human condition where she gained firsthand perspectives on worldwide needs, values and challenges.

The issues of "what if," "why," and "how" dominated her search to understand the broader perspectives of world strategies, cultural norms, and expectation of how to prepare individuals, families, and communities for being in the world, surviving and thriving. We are not alone and need each other standing as equals became her trademark and motto in search of how to prepare educators in different cultures regarding the blending cultural needs with innovative strategies and techniques of national heroes of health and healing with native traditions. She stood alongside those she served as equals not as "I am better than," or "have all the answers," to a "we are one," "united by our humanness as equals," and "each with different perspective on the meaning of life," living in this world in a cooperative way with a win-win perspective as her personal guiding light which serves as the philosophy in her new role leading Akamai University as their first woman president and innovator holding the light for the faculty and students to shine their light.

Medani P. Bhandari holds M.A. Anthropology (Tribhuwan University, Nepal), M.Sc. Environmental System Monitoring and Analysis (ITC-The University of Twente, the Netherlands), M.A. Sustainable International Development (Brandeis University, Massachusetts, USA), M.A. and Ph.D. in Sociology (Syracuse University, NY, USA).

Prof. Bhandari is a well-known humanitarian, author, editor, co-editor of several books and author of hundreds of scholarly papers on social and environmental sciences: a poet, essayist, environment, and social activist, etc. He is a true educator, who advocates that education should not be only for education shake and for employment (get position, earn money, etc.) but should bring meaningful change in students' lives, which could provide the inner peace in life and be able to face any kind of challenges. Prof. Bhandari is also a motivational speaker—life skill coach (how to remain in peace and calm when situation is not in our control, with application of Basudhaiva Kutumbakam principles)—and in writings and teaching, he exemplifies how studies of scientific theories can be enjoyable and applicable in day-to-day life. As such the epistemology of knowledge which is the basis of theory building is considered not fun subject for everybody; however, Prof. Bhandari's works show that, there is always lucid and attractive side in difficult subject.

Dr. Medani P. Bhandari has been serving as a Professor of inter-disciplinary Department of Natural Resource & Environment / Sustainability Studies, at the Akamai University, USA and Professor at the Department of Finance, Innovation and Entrepreneurship, Sumy State University (SSU), Ukraine. Dr. Bhandari is also serving as Editor in Chief of International Journal—The Strategic Planning for Energy and the Environment (SPEE), Denmark, The Netherlands, USA- ISSN: 1546-0126 (Online Version), ISSN: 1048-4236 (Print Version) https://www.journal.riverpublishers.com/index.php/SPEE/about He is also serving as managing editor at the Asia Environment Daily, (the largest environmental journal of the world)- https://asiaenvdaily.com/index.php/editorials-2/2-uncategorised/29760-about-asia-environmental-daily

Dr. Bhandari has spent most of his career focusing on the Social Innovation, Sociological Theories; Environmental Sustainability; Social Inclusion, Climate Change Mitigation and Adaptation; Environmental Health Hazard; Environmental Management; Social Innovation; Developing along the way expertise in Global and International Environmental Politics, Environmental Institutions and Natural Resources Governance; Climate Change Policy and Implementation, Environmental Justice, Sustainable Development; Theory of Natural Resources Governance; Impact Evaluation of Rural Livelihood; International Organizations; Public/ Social Policy; The Non-Profit Sector; Low Carbon Mechanism; Good Governance; Climate Adaptation; REDD Plus; Carbon Financing; Green Economy and Renewable Energy; Nature, Culture and Power.

Dr. Bhandari's major teaching and research specialties include: Sociological Theories and Practices; Environmental Health; Research Methods; Social and Environmental innovation; Social and Environmental policies; Climate Change Mitigation and Adaptation; International environmental governance; Green Economy; Sustainability and assessment of the economic, social and environmental impacts on society and nature. In brief, Prof. Bhandari has sound theoretical and practical knowledge in social science and environment science. His field experience spans across Asia, Africa, the North America, Western Europe, Australia, Japan, and the Middle East. Dr. Bhandari has published several books, and scholarly papers in international scientific journals. His recent books are *Green Web-II: Standards and Perspectives from the IUCN (2018); 2nd Edition 2020; Getting the Climate Science Facts Right: The Role of the IPCC; Reducing Inequalities Towards Sustainable Development Goals: Multilevel Approach; and Educational Transformation, Economic Inequality—Trends, Traps and Trade-offs; Social Inequality as a Global Challenge, the Unbeatable Challenges of Inequality,*

Perspectives on Sociological Theories, Methodological Debates and Organizational Sociology. Additionally, in creative writing, Prof. Bhandari has published 100s of poems, essays as well as published two volumes of poetry with Prajita Bhandari.

Natalia V. Eremeeva has a Ph.D. in Law, Institute of Economic and Legal Research of NAS of Ukraine, Head of the Department of Economic and Legal Studies of Economic Security and Safety Issues. She carries out scientific and applied research, scientific expertise, and counseling in the sphere of legal support of the economy, legal regulation of economic activity. The main research interests are economic law, modernization of financial legislation, legal support of energy safety.

Oksana Ponomarenko graduated from Sumy National Agrarian University (M.S in "Management of foreign economic activities"). Currently, she is a PhD student (Sumy State University) in "Economics." She worked as a Research Assistant of the Department of foreign economic relations at Sumy State University (Ukraine); participated in research groups conducted grant research funded by governments of Ukraine and the Czech Republic; had an experience of working in bank sphere and international IT Company. She is the author of 12 papers. Scientific interests include sustainable development, local finances, socio-economic development, local communities, cooperatives, management, and regional development.

Olena Pavlenko graduated from Sumy State University (M.S in «Management of Organizations», Ph.D. in "Environmental Economics and Natural Resources Protection"). Currently works as an Associate Professor at the Department of Management at Sumy State University (Ukraine). She took part in Tempus Projects INARM «Eastern-European Qualification Framework in The Field Of Informatics And Management» and ALIGN «Achieving and Checking the Alignment between Academic Programmes and Qualification Frameworks» (2012–2015). She is a member of Ukrainian Association for the Development of Management and Business Education (UARMBO). She is the author of more than 80 papers. Scientific interests include regional management, management of social and environmental safety, Operations Management. She is the Leader of the working project group (Program Leader) of the «Civil Service» educational program.

Yuriy Petrushenko graduated from Sumy State University (M.S in Finance, 1999). In 2003, he graduated from V. N. Karazin Kharkiv National University. Ph.D. in Economics (2004), Associate Professor (2006), Doctor of Economics (2016), Professor (2018). Currently works as a Head of International Economic Relations at Sumy State University (Ukraine). Scientific interests include lifelong learning, international communications, the effectiveness of international projects and programs, economic aspects of globalization, social and public finances. He received research grants from such international organizations as the National Council of Economic Education (USA), the Center for Advanced Studies and Education (CASE, Lithuania), The Economics Education and Research Consortium (EERS), and the United Nations Development Program (UNDP Ukraine), Swiss Federal Department of Foreign Affairs (FDFA), Bundeszentrale für politische Bildung (BpB, Germany). He has more than ten years of experience in managing international projects. He was the coordinator of the UN Development Program partner universities network in Ukraine during 2011–2013. Since 2017—GIZ expert, in 2017–2019, he worked as an advisor on financial decentralization in the project U-LEAD with Europe: Ukraine—Local Empowerment, Accountability, and Development Program. Since 2015, he has been constantly involved in educational and research projects of the Institute for International Cooperation of the German Adult Education Association (DVV International).

Pakhnenko Olena was graduated from the State Higher Educational Institution "Ukrainian Academy of Banking of the National Bank of Ukraine," obtained a master's degree in Finance (2009). She gained Ph.D. in Economics in specialty 08.00.08—Money, Finance, and Credit (2012). Associate Professor, Department of Finance, Banking and Insurance, Sumy State University. Research interests: financial inclusiveness, financial literacy, financial market, digital finance.

Prem B. Bhandari is a Social Researcher (Demographer). He has a Ph.D. in Rural Sociology and Demography from the Pennsylvania State University, USA. He completed his M.Sc. in Rural Development Planning from the Asian Institute of Technology, Thailand and earned a B.Sc. degree in Agricultural Economics from Tribhuvan University, Nepal. Currently, Bhandari is working as the Evaluation Coordinator/Data Analyst for a School Feeding Program Evaluation in Cambodia and Haiti through KonTerra Research Group, Washington, D.C. In addition, he is the Managing Director of South Asia Research Consult, Inc. He is also an Adjunct Professor at Agriculture and Forestry University in Nepal. Prior to this, he worked as a researcher (Co-Investigator) and Assistant Research Scientist at the Institute for Social Research, University of Michigan. He has over 20 years of experience in designing and managing large scale social (survey) research programs. In Nepal, he worked as the assistant professor of agricultural economics at Tribhuvan University, Nepal. His major research interests include social research methods (survey design, survey instrument design, pre-testing and finalization, survey data collection, survey management and quantitative analysis of survey data, and qualitative surveys); socio-economic and cultural inequalities and determinants of demographic behaviors (population health and migration); social inequalities; food security and nutrition; and overall rural social change. He has a strong background and experience in quantitative (and qualitative) analysis. Bhandari has published a number of articles in peer-reviewed journals and as book chapters and presented dozens of papers in national and international professional meetings. He is the editorial member of several journals and an occasional reviewer of over a dozen of peer-reviewed journals.

Rubanov Pavlo is Associate Professor, Department of Finance and Entrepreneurship, Sumy State University Ph.D. in Economics (08.08.01—Economics of nature use and environment protection, 1998) Ph.D. thesis: "Efficiency of the mechanism of balanced development of environmental and economic interests," Research interests: alternative financing of business based on online platforms; regulation of activities of financial Intermediaries; financial activity, and capital structure of business entities; finance of sustainable development. **Carlos Sánchez** is Ph.D. candidate in Economics from the National University of Mexico. National Research Award by National University of Mexico, PricewaterhouseCoopers Mexico, and Interacciones Financial Group in 2014. Member of the Young Sinologists Program of Shanghai Academy of Social Sciences and the World Association of Political Economy. Author of "Knowledge Capitalism and State Theory"—a space-time approach explaining development outcomes in the Global Economy." He is a Professor of International Economics, World Economic Structure, and Economic Integration at the National University of Mexico. His research topics are knowledge economy, state, poverty, and inequality.

Sergii Polumiienko graduated from the Probability Theory and Mathematical Statistics Department of the Kyiv State Taras Shevchenko University, Doctor in Physics and Mathematics. Specialized in game theory and its applications, heuristic algorithms for decision making and complex applied systems modeling. Author of about 130 scientific papers and 10 monographs. Among the results—one of the authors (head) of the adopted by the Cabinet of Ministers of Ukraine National system of indicators of the information society development level. Now—Head of Department of the Institute of Telecommunications and Global Information Space of the National Academy of Sciences of Ukraine, Chokolivsky Blvd., 13, 03186, Kyiv, Ukraine. Before he was Deputy Director of the Institute of Applied Informatics, JSC.

Serhiy Lieonov is vice-rector of Sumy State University. He graduated from the Sumy State University in 1999. He got a PhD in speciality 08.02.02—Economics and Management of Scientific and Technical Progress (2003), Associate Professor of Management Department (2006), Doctor of Economics in speciality 08.00.08—Money, Finance and Credit (2010), Professor of Finance Department (2012). Since 2017, he is Professor at Economic Cybernetics Department. Professor Lieonov is a research leader at his department. Research interests are in the spheres of banking, investing, reproductive processes in the economy, risk management, innovation, corporate reporting and its audit, investment activity. He is the author of more than 150 scientific papers and more than 50 monographs and numerous presentations. Professor Lieonov is a Head of the Specialized Academic Committee at Sumy State University; he is a mentor for many Ph.D students and Post-Docs. He is a contributor and a member of the Working Group of scientists who develop the forecasting of economic and social development of the Sumy region for the period 2018–2022. He is an active participant in the social projects for the local community, as well as educational events.

Tetyana Vasilyeva is a Director of Academic and Research Institute of Business, Economics and Management, Professor (2010, Professor's degree of the Department of Management) and Doctor of Economics (2008, speciality 08.00.08—money, finance, and credit). She is the author of more than 200 scientific papers, 70 monographs, and numerous materials of the conferences. She is Editor-in-chief of the scientific journal "Business Ethics and Leadership," member of the editorial board of the journals: "Strategy and Development Review," "Carbon Accounting and Business Innovation Journal." She is a mentor for PhD students and Post-Docs and is a member of the Specialized Academic Committee at Sumy State University; one of the founders and a member of NGO "Council of Young Scientists." She was the initiator of launching the Business Support Center in Sumy region, and now she is a member of this organization that gained the support of the European Bank for Reconstruction and Development. She is a team leader, an author, and participant of numerous scientific projects that gained the highest scores in Ukraine. Research interests include banking, investment, risk management, innovation activity, social responsibility, education. She is a leader and contributor in international projects, such as Jean Monnet projects (2014, 2015, 2017), grant projects such as "Development of Dialogue Between Banks and Civil Society in the Context of Ensuring Democratic Processes in Ukraine" (DAAD, 2016), "Enhancing Energy Security by Swiss–Ukrainian–Estonian Institutional Partnership" SCOPES IZ74Z0_160564 (Swiss National Science Foundation, 2015–2017) and many others. She is a contributor in many educational projects financed by Ministry of Economic Cooperation and Development of Germany, DVV International, DESPRO in Switzerland, American Council for Economic Education NCEE, International Foundation "Vidrodzennia" etc. Professor Vasilyeva initiated many scientific and educational events concerning the development of corporate social responsibility (e.g., winter/summer schools for youth "Economic and business values," "School of innovations and social entrepreneurship," "Management of socially responsible business," etc.)

Yuliia Matvieieva graduated from Sumy State University with a master's degree in «Management of Organizations» and a Ph.D. in «Environmental Economics and Natural Resources Protection»). The topic of the dissertation—«Ecologically oriented management of administrative-territorial units' socio-economic development». Grant activities: The Ukrainian-France R&D joint project for the period of 2015–2016. Topic: «The Recovery of Sustainable Competitiveness of National Economy: Ecological Imperative and Corporate Responsibility». Currently, she works as a Senior Lecturer at the Oleg Balatskyi Department of Management, Sumy State University (Ukraine). She is a Ukrainian Association member for the Development of Management and Business Education (UARMBO) from 2020. Her research focuses on diagnosing social, economic, and ecological systems; smart Grid-technology; inequality, and socio-economic drivers. She is the author of more than 85 papers.

Yuliia Opanasiuk graduated from Sumy State University with a master's degree in «Economy of the enterprise» and a PhD in «Environmental Economics and Natural Resources Protection»). The topic of the dissertation—«Ecological and economic evaluation of man-made disasters». Grant activities: «Business security: national and global aspects» (Information System Management University, Riga, Latvia, 2019), «Health recreation in urban agglomerations (within the CSF achievements in Ukraine)» (grant of the President of Ukraine—responsible executor, 2019).

Currently, she works as a Senior Lecturer at the Oleg Balatskyi Department of Management, Sumy State University (Ukraine). She is a Ukrainian Association member for the Development of Management and Business Education (UARMBO) from 2020. Her research focuses on Economy disasters, Public administration. She is the author of more than 60 papers. More information at https://management.fem.sumdu.edu.ua/en/opanasiuk

Yuliia Opanasyk, is a senior lecturer at Oleg Balatskyi Department of Management of Sumy State University. She gets her PhD in economics. She is the author of numerous publications in the field of management.

For Product Safety Concerns and Information please contact our EU representative GPSR@taylorandfrancis.com Taylor & Francis Verlag GmbH, Kaufingerstraße 24, 80331 München, Germany